XXI Meeting of Physics 2021 (MOPHY XXI 2021)

Journal of Physics: Conference Series Volume 3052

Cusco, Peru
16-18 December 2021

Editors:

Sayri T. G. Roca
Carlos J. S. Salinas
Milton R. Gamarra

ISBN: 979-8-3313-2479-7
ISSN: 1742-6588

Printed from e-media with permission by:

Curran Associates, Inc.
57 Morehouse Lane
Red Hook, NY 12571

Some format issues inherent in the e-media version may also appear in this print version.

Additional copies of this publication are available from:

Curran Associates, Inc.
57 Morehouse Lane
Red Hook, NY 12571 USA
Phone: 845-758-0400
Fax: 845-758-2633
Email: curran@proceedings.com
Web: www.proceedings.com

TABLE OF CONTENTS

Journal of Physics: Conference Series **3052** (2025) 011001 doi:10.1088/1742-6596/3052/1/011001

Proceedings of the XXI Meeting of Physics

Cusco, December 16-18, 2021

Universidad Nacional de San Antonio Abad del Cusco

Meeting of PHYSICS

Editors

Sayri Tupac GARCIA ROCA (UNSAAC, Peru)

Carlos Javier SOLANO SALINAS (UNMSM, Peru)

Milton ROJAS GAMARRA (UNSAAC, Peru)

Journal of Physics: Conference Series **3052** (2025) 011001 doi:10.1088/1742-6596/3052/1/011001

PREFACE

The ***Meeting of Physics*** is an event that aims to motivate and promote Physics in Peru and in the region, presenting the scientific contributions of researchers and students from different universities and research centers from Peru and abroad. This event is also a great opportunity to meet and initiate scientific collaborations with local and international guests some of these are renowned scientists.

Unfortunately, during 2020 and 2021 the global pandemic that affected us all also drastically slowed research activities. Despite all these, we managed to organize this *Meeting of Physics* in a hybrid mode. It took a while for researchers and students to complete their articles for the proceedings and for us to get the corresponding funding, but finally this ***Proceedings of the XXI Meeting of Physics*** were produced.

In this *Proceeding* we present a selection of works that were exhibited in the plenary and parallel lectures, as well as in "poster" sessions.

The editors wish to thank the Faculty of Chemical, Physical and Mathematical Sciences of the **Universidad Nacional de San Antonio Abad del Cusco (UNSAAC)** in Cusco, Peru, and all those who made the *XXI Meeting of Physics* possible.

The editorial Committee:

Sayri Tupac GARCIA ROCA (Universidad Nacional de San Antonio Abad del Cusco, Peru)

Carlos Javier SOLANO SALINAS (Universidad Nacional Mayor de San Marcos, Peru)

Milton ROJAS GAMARRA (Universidad Nacional de San Antonio Abad del Cusco, Peru)

Journal of Physics: Conference Series **3052** (2025) 011002 doi:10.1088/1742-6596/3052/1/011002

Peer Review Statement

All papers published in this volume have been reviewed through processes administered by the Editors. Reviews were conducted by expert referees to the professional and scientific standards expected of a proceedings journal published by IOP Publishing.

- **Type of peer review:** Double Anonymous
- **Conference submission management system:** Morressier
- **Number of submissions received:** 8
- **Number of submissions sent for review:** 8
- **Number of submissions accepted:** 7
- **Acceptance Rate (Submissions Accepted / Submissions Received × 100):** 87.5
- **Average number of reviews per paper:** 1.95
- **Total number of reviewers involved:** 6
- **Contact person for queries:**
  Name: Carlos Javier Solano Salinas
  Email: csolanos@unmsm.edu.pe
  Affiliation: Universidad Nacional Mayor de San Marcos

Journal of Physics: Conference Series 3052 (2025) 012001 doi:10.1088/1742-6596/3052/1/012001

Correlation between shear near-field acoustic emission and the radius of curvature of a water bridge meniscus formed between parallel plates

Cesar E Sotelo[1], Daniel F Reyes[1], Ruben E Bringas[1], Kacharat Supichayanggoon[2], Carlos J Solano[3], and Andres H La Rosa[1,2]
[1]*National University of Engineering, Physics Department, Lima, Peru*
[2]*Physics Department, Portland State University, P.O. Box 751, Portland, Oregon 97207, USA*
[3]*Universidad Nacional Mayor de San Marcos, Faculty of Physical Sciences, Lima, Peru*
Email: andres@pdx.edu

Abstract. A water bridge, formed through capillary condensation at the nanometer-sized gap defined by a sharp probe and a flat sample, produces acoustic emission when subjected to shear motion. Very recently, it has been found that such an emission is sensitively detected in the near-field zone by the Shear-force Near-field Acoustic Microscope (SANM). When additionally subjected to vertical elongation and compression, reports from experiments performed in an AFM set up and complemented with a corresponding hydrodynamic model, indicate that the meniscus remains in mechanical equilibrium (but not in thermodynamic equilibrium) while maintaining its volume constant. Herein, we present numerical solutions of the hydrodynamic equation governing the profile of the meniscus formed between two parallel plates, and verify calculations previously reported for similar confining geometries. Significantly new, we find a correlation between the acoustic emission obtained in SANM experiments and the inverse of the meniscus' total radius of curvature at different gap distances between the plates. While the fitting may not be universal (as the experiment depends on different parameters) the trend of agreement between the model and experimental results is encouraging. These measurements are relevant to investigations on the dynamic behavior of fluids near hydrophobic/hydrophilic substrates, wetting properties of solids, and nanotribology phenomena in general.

1. Introduction

Water menisci form inside porous materials or around sharp asperities, below saturation conditions, driven by capillary condensation, [1, 2] a phenomenon that is relevant to many areas, namely: nanolithography [3-6], nanotribology [7, 8], corrosion [9, 10], and the fabrication of [11].) Altogether, nanofabrication and characterization processes have indeed triggered numerous efforts to understand systems involving nanoscale liquid bridges and confined fluids in general.

A specific aspect addressed in this manuscript, and still not yet well understood, is the role played by confined fluids in the acquisition of high-resolution scanning probe microscopy (SPM) images, namely Atomic Force Microscopy (AFM) and Near-field Scanning Optical Microscopy (NSOM), where a water meniscus bridge (that fills the gap between a SPM sharp probe and a sample) accompanies the probe's apex during the imaging raster scan. In particular, there is an interest for visualizing the morphology of the meniscus *i*) to verify whether the meniscus' curvature is compatible either with the Kelvin's radius (a *thermodynamic equilibrium* state parameter) or with simpler *mechanical equilibrium*

Journal of Physics: Conference Series **3052** (2025) 012001 doi:10.1088/1742-6596/3052/1/012001

states; as well as *ii*) to help identifying specific meniscus' morphology parameters (curvature, length, area, etc.) that could be correlated with the meniscus' viscoelastic response (this second aspect, pertinent to this paper, is described in more detail below). In short, the detailed morphology of the meniscus bridging the probe's apex and the sample acquires particular importance. However, given the meniscus nanometer-sized dimension, its visualization is out of reach for conventional optical microscopy due to its inherent limited spatial resolution ~ $\lambda/2$ (where λ is the wavelength of the light used; typically λ = 500 nm) imposed by far-field diffraction effects. As an alternative, "Environmental SEM" [12] [13] has overcome this shortcoming by exploiting its high-resolution capability even when operating at close to environment conditions. In effect, environmental SEM has been used to visualize de sudden formation of a water meniscus between a tapered solid probe and a solid flat sample when placed at nanometer-sized separation distances (thus corroborating its predicted stochastic formation). But, despite this accomplishment, still there is a long way to fully understand the dynamic behavior of a nanoscopic-sized water meniscus. For example, what properties are involved in the (elastic and inelastic) energy dissipation channels when the probe interacts with a surface? To address this shortcoming, analytical or numerical solutions of the corresponding hydrodynamic models describing the dynamics of a water meniscus are required.

On the other hand, the relatively newer Shear-force Near-field Acoustic Microscopy (SANM) [14-16] technique has demonstrated capability to detect the near-field acoustic emission from the water meniscus trapped between a sharp probe and a flat sample while under shear motion. This acoustic signal has become a primary candidate to account for the elastic energy dissipation channel in probe-sample interfacial interactions [2]. AFM measurements have also revealed important characteristics of these interactions, like, for example, the preservation of the water meniscus volume (that bridges the probe and the surface) when subjected to vertical compression-stretching cyclic process [1]. Such a result has shed light on the type of solutions (that render the morphology of the meniscus) to be admitted from the corresponding hydrodynamic equation (as further addressed in the sections below)] and thus guide better the search for meaningful meniscus' (physical, geometrical) parameters that control its near-field acoustic emission, as well as its dynamic response to interfacial interactions in general, which is precisely what we plan to address in this manuscript.

In the first part we provide a brief description of the SANM technique, addressing *i*) how the water meniscus bridge is formed, *ii*) how the probe is purposely driven laterally (a strategy serving the purpose to figure out how close is the probe to the sample), and *iii*) how our group capitalized this setting to sensitively detect the acoustic waves emitted by the water meniscus. Next, we address the numerical solution of the hydrodynamic Eq (1) (which predicts the morphology of the water meniscus) via a 4th order Runge Kutta method. To simplify this latter task, we consider two parallel plates as the boundary constrains of the eventual formation of the meniscus. The meniscus' profiles are obtained for different separation distances "d" between the plates. Subsequently, several parameters "p" (associated to each meniscus profile) are intuitively identified and plotted as a function of "d". Finally, attempts are performed to correlate the numerically calculated "p vs d" traces with the "*near-field acoustic emission* vs d" traces obtained from SANM experiments. Overall, this work represents a first attempt towards identifying the physical properties and geometrical morphology of the meniscus that are responsible for the elastic component of the energy dissipation in interfacial probe-sample shear interactions.

2. The SANM apparatus

SANM is used to investigate, via acoustic sensing [14-16], the viscoelastic properties of fluids trapped between the apex of a tapered probe and a flat substrate [17,18]. Here we provide a succinct description [14,19].

2.1. Stochastic formation of a water meniscus

When a sharp probe approaches a flat substrate the spontaneous formation of a fluid bridge occurs (Fig.1). Depending on the environmental humidity conditions, probe's geometry, and substrate

Journal of Physics: Conference Series **3052** (2025) 012001 doi:10.1088/1742-6596/3052/1/012001

materials (hydrophobic or hydrophilic), a meniscus forms stochastically anywhere in the 2 nm to ~50 nm range of the probe-substrate separation distance [2, 20].

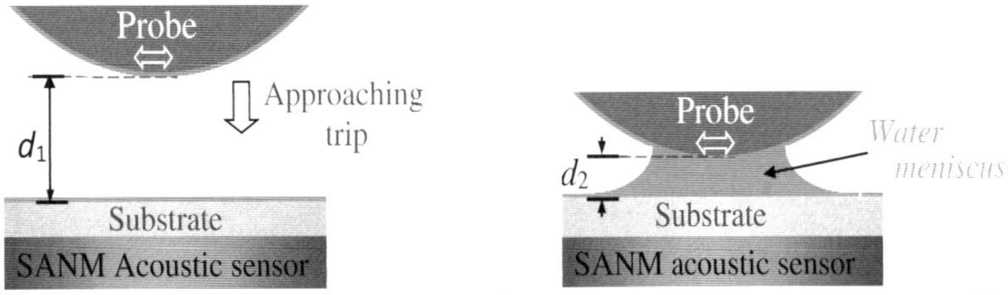

Figure 1. Left: Laterally oscillating sharp probe approaching a flat sample**. Right:** At ambient conditions, when the probe is placed at nanometer-sized distance (30 nm or lower) from the sample, a water meniscus is stochastically formed.

2.2. *The probe*
The SANM analytical system is shown in Fig.2. A fiber glass (~1 mm long, 125 μm diameter) with an apex of conical shape is attached to one leg of a 32 kHz quartz tuning fork (QTF). From now on, the QTF together with its attached tapered fiber will be referred to as *the probe*. The whole probe is then attached to a translation stage (Nano-Z50HS, MCL S02472) whose vertical position is controlled by applying a DC voltage.

2.3 Synchronous detection of the probe-sample distance
To control the probe-sample distance a synchronous detection strategy is employed. To that effect, first the QTF is electrically driven (~ 4 mV ac), which causes the tines to oscillate laterally (~10 nm amplitude). As the probe is lowered towards the sample (using the Nano-Z50HS), it is observed that the lateral amplitude decreases monotonically in the last ~20 nm before the apex touches the sample. The decrease in amplitude is attributed to the interaction of the probe with the water meniscus that forms between the apex and the sample, as alluded in Section 2.2 above. Fig.2 shows a synchronous detection (using the lock in amplifier #1) of the probe's lateral oscillations amplitude. The deeper the probe gets immersed intro the meniscus, the lower its amplitude of oscillations; the amplitude of the lateral oscillations serves then as an indicator of how close the probe's apex is to the surface.

2.4. *Detection of the near-field acoustic signal*
As the probe's apex laterally oscillates inside the water meniscus, an acoustic wave (the SANM signal) in engendered, which couples to the sample (a silicon oxide wafer, for example) and subsequently reaches the acoustic sensor (VS30-V AE sensor; Acoustic Technology Group, Michigan, USA) located in intimate contact with the back side of the sample. Since the acoustic transducer is located at a distance equal to the thickness of the sample (~ 0.2 cm) much smaller than the wavelength of the operating acoustic wave ($\lambda_{acoustic}$ ~ 20 cm in silicon material and $\lambda_{acoustic}$ ~5 cm in bulk water) [21] the detection is in the realm of the "near-field" region. Fig.2 shows the SANM synchronous detection via the lock in #2.

Journal of Physics: Conference Series **3052** (2025) 012001
doi:10.1088/1742-6596/3052/1/012001

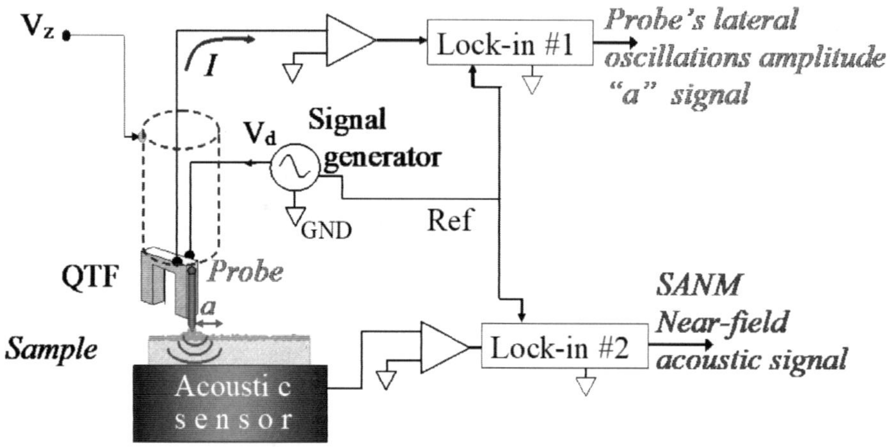

Figure 2. Schematic of the experimental setup of the Shear-force Near-field Acoustic Microscope (SANM). The central part shows a tapered glass fiber attached to an electrically driven quartz tuning fork. The stylus undergoes lateral oscillations of amplitude "*a*", whose value (of few nanometers) decreases as the apex vertically approaches the sample. The top side shows the synchronous detection of the probe's lateral oscillations (using the lock-in #1). The bottom side displays the synchronous detection of the near-field acoustic emission from the fluid (using the lock-in #2), which typically increases as the probe approaches the sample.

3. Numerical solutions to obtain the morphology of the water meniscus

In this section, we pursue solving the profile of the meniscus that forms between two planar surfaces using a system of differential equations derived from the Young-Laplace equation [22]. The system describes the shape of the meniscus as a function of the contact angle, surface tension, and curvature radius between two surfaces.

.

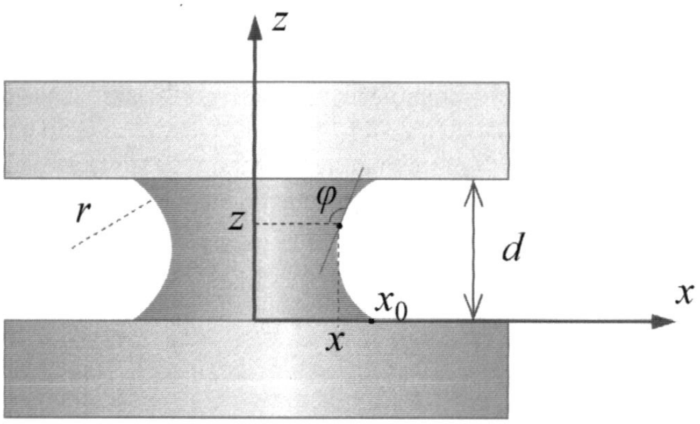

Figure 3. Meniscus profile between two planar surfaces

The equations for the horizontal and vertical positions of the meniscus can be expressed as [22],

$$\frac{dx}{d\varphi} = \frac{\cos(\varphi)}{\frac{1}{r} - \frac{\sin(\varphi)}{x}}$$
$$\frac{dz}{d\varphi} = \frac{-\sin(\varphi)}{\frac{1}{r} - \frac{\sin(\varphi)}{x}} \tag{1}$$

where x is the horizontal distance; z the vertical distance, φ the angle, and r is the radius of curvature.

The system is numerically solved using the fourth-order Runge-Kutta method [23], which integrates the differential equations governing the meniscus profile. To generate the initial profile, specific values for the radius of curvature (r) and the x_0 coordinate are selected, the latter representing the point where the meniscus contacts the surface. Subsequently, the two planar surfaces are brought closer together, causing the meniscus to change its shape while enforcing the condition to maintain a constant volume (the latter requirement based on experimental results from a similar experiment [1]). To fulfill this latter requirement, a bunch of profiles are obtained by systematically varying both the radius of curvature (r) and the initial contact position (x_0). Each profile is examined to ensure that it reflects the imposed physical constraints of the system such as volume conservation and contact angle with the surfaces.

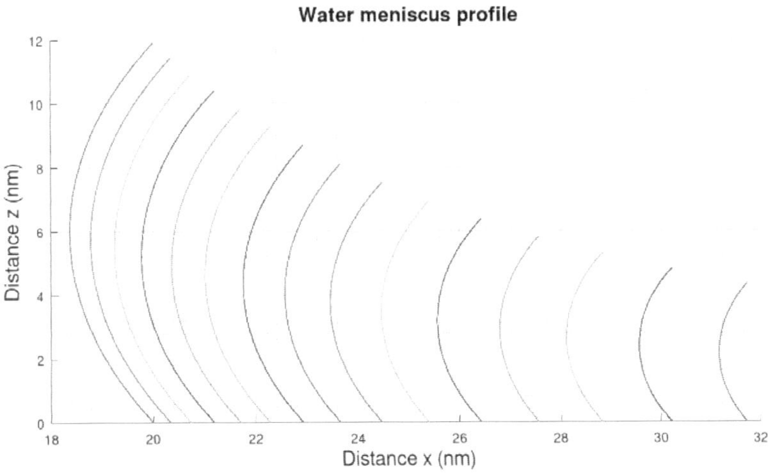

Figure 4. Numerical solutions of Eq. (1). Profile of water menisci formed between two parallel plates, assuming a contact angle of 60 degrees in both plates. The profiles correspond to different vertical separation distances between the plates, while maintaining the water meniscus' volume constant.

One of the primary difficulties encountered during the simulation is precisely enforcing a constant volume through the iterations. Since the equation depends on radius of curvature (r), a slight variation in r can lead to significant change in the meniscus shape and volume. Therefore, it became necessary to accurately control the values of r and x to ensure that the calculated volume does not deviate beyond a specified tolerance from the initial volume. Finding such correct values poses a computational challenge. A too simple and direct approach, where all possible values of x_0 are tested within a range, turns out to be inefficient and slow, especially when a high level of accuracy is required. To address this shortcoming, we resorted to numerical optimization techniques such as binary search to efficiently narrow down the values of x_0 that minimize the volume error. Accordingly, we first use a reference meniscus profile for a given initial separation distance between the constraining parallel surfaces. This profile serves as a baseline for determining the subsequent meniscus shapes as the surfaces approach each other. At each step, we adjust the value of x_0 by using a binary search, ensuring that the volume remains within a 0.05% tolerance of the reference volume. By iteratively refining the values of x_0 and updating the initial conditions for the next step, we can efficiently generate meniscus profiles for different surface separations while keeping the volume constant. This approach not only accelerates the computational process but also maintains the required accuracy in volume preservation. Figure 4 shows the water meniscus profile for different values of the radius of curvature and different x_0 values. Each profile has the same volume within the 0.05% tolerance.

4. Interpretation

The numerical solutions obtained in the previous section offer an opportunity to investigate what geometrical parameters of the meniscus play an important role in the emission of the near-field acoustic wave. This inquietude springs from the fact that, as observed in Fig.5 (diagram in the right side), when the probe approaches the sample one observes that the acoustic signal increases despite the corresponding monotonic decrease of the probe's amplitude of lateral oscillations. One would expect that both signals, the acoustic emission and the probe's lateral oscillations, to maintain some kind of proportional relationship (like the louder a person intends to talk, the louder his/her voice should sound). The opposite occurs instead in the SANM experiments. Could it be that the deeper the probe gets immersed into the meniscus, the denser the water encountered by the apex and, hence, being able to produce louder sound? We have not found in the literature evidence to sustain the hypothesis that the density of water layers increases dramatically in the last 20 nm region closer to a substrate.

On the other hand, being an acoustic wave the result of pressure variations in a fluid, a parameter associated to pressure variation in the meniscus is the Laplace pressure $P_L = \sigma/r$, where σ is water surface tension and r the meniscus' radius of curvature. Accordingly, maybe the acoustic emission correlates with the Laplace pressure? To evaluate this conjecture, we must first find out how σ/r varies with the distance d between the constraining parallel plates. The results shown above in Fig. 4 come then handy to explore such a dependence, which is shown in Fig. 5 (diagram on the left side).

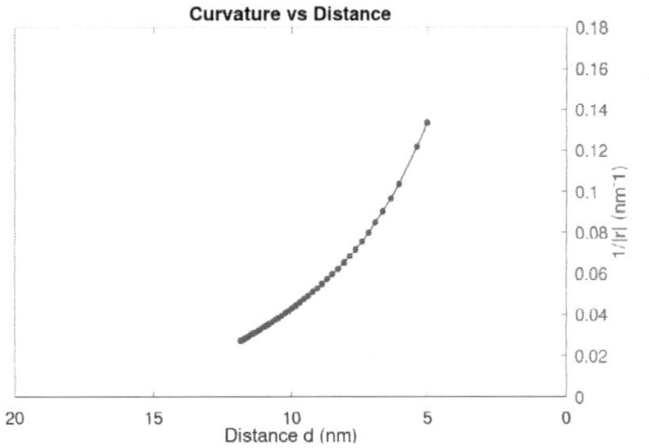

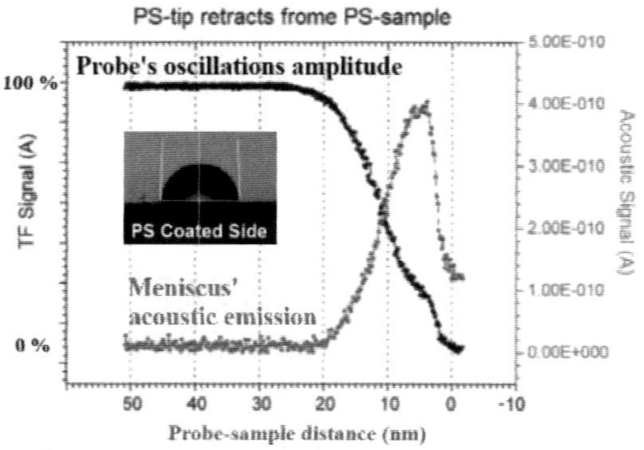

Figure 5. Left: Calculated meniscus' curvature $1/r$ vs distance d (vertical separation between the boundary plates) obtained from solving numerically Eq (1). **Right:** Experimental SANM results (probe's amplitude of lateral oscillations and acoustic emission as a function of the probe-sample distance) obtained using a polystyrene coated tapered probe and a polystyrene (PS) coated flat substrate. The inset shows the contact angle a drop of water makes with the substrate, which displays that PS is hydrophobic.

The monotonic behavior and the positive concavity of the (calculated) $1/r$ trace shown on the left side diagram of Fig.5 is encouraging, since both characteristics coincide with typical behavior of the experimentally measured SANM signals. To verify our conjecture, we have selected SANM experimental data that, even though comes from an experimental setting somewhat different (a sharp

Journal of Physics: Conference Series **3052** (2025) 012001 doi:10.1088/1742-6596/3052/1/012001

probe, instead of a parallel plate, approaches a flat sample) it should keep the main characteristics accompanying the acoustic emission of a meniscus. More specifically, the probe is a taper optical fiber coated with polystyrene (PS) (contact angle = 90°, as shown by the inset in the right side of Fig. 5) and the sample is a silicon wafer substrate coated also with PS. Before implementing a comparison, the acoustic trace (shown on the right side of Fig.5) must be normalized, so to obtain an acoustic trace associated to a case in which the probe oscillates with the same amplitude while the apex approaches the sample. Such a normalization is obtained dividing the raw acoustic trace by the probe's amplitude trace. The (experimental) normalized acoustic wave and the (calculated) Laplace pressure are both plotted together in Fig.6. Since the acoustic signal is in fact the electrical output signal from the acoustic sensor, a scaling factor was selected to obtain the correlation thus displayed by these two signals. The close matching correlation between the two traces is gratifying.

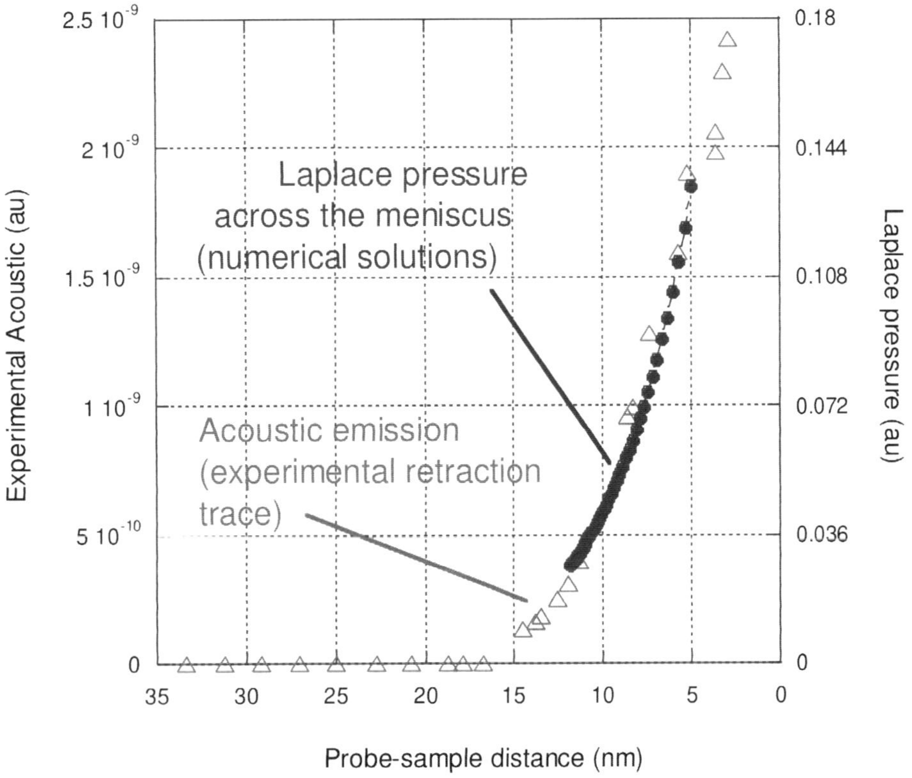

Figure 6. Comparison between the (experimentally measured) near-field acoustic wave and the (calculated) Laplace pressure across the meniscus. A scaling factor was arbitrary selected to obtain the shown correlation between these two signals.

5. Conclusions

The search for an explanation of the origin of shear forces (those interactions that slow down the lateral oscillation of a tapered probe, as used in scanning probe microscopy) has a long standing in time since the invention of the atomic force microscope and the near field optical microscope during the 90's. The monotonic behavior of the probe's lateral oscillations observed when the probe gets near a flat surface in the last ~ 30 nm region has given rise to many conjectures along all these years, all with valid arguments and accompanied contradictions. Questions about the origin of shear forces in the context of interfacial friction by Karrai in the year 2000 [24] were placed in very sounding grounds (measurements were made under 10^{-5} Torr vacuum and the shear forces were still sensed), which implied the need of new hypothesis to reach a plausible explanation about the origin of shear forces. This manuscript has introduced for the first time a plausible explanation, backed with experimental data, on the near-field acoustic emission as an elastic energy dissipation channel in shear force interactions. The origin has been identified in the variations of the meniscus' curvature as the probe approaches the sample. The result in Fig.6 is the best evidence we have been able to provide. More systematic corroboration, with simulation using different contact angle and different boundary plates geometries are underway.

Acknowledgment
ALR acknowledges the hospitality provided by the National University of Engineering during the preparation of this manuscript in Lima-Peru. The lengthy and direct fruitful discussions with the local Peruvian scientists, in particular the participants within the "Programa de Doctorado en Ciencias con Mención en Física," is greatly appreciated. Gratitude is also extended to Portland State University for granting ALR the opportunity to fulfill his sabbatical research project at National University of Engineering Lima-Peru.

References

[1] Sirghi L, Szoszkiewicz R and Riedo E 2006 Volume of a nanoscale water bridge *Langmuir* **22** 1093-1098
[2] Wang X, Fernandez R, Brockman T, Supichayanggoon T, La Rosa A H 2024 Monitoring water meniscus formation at nanocontacts with shear-force acousto near-field microscopy *J. Appl. Phys.* **135** 244702. https://doi.org/10.1063/5.0215054
[3] Walker D A, Hedrick J L, Mirkin C A 2019 Rapid large-volume thermally controlled 3D printing using a mobile liquid interface *Science* **366** 360 doi: 10.1126/science.aax1562
[4] Wang Xiaohua, Wang Xin, Fernandez R, Yang M and La Rosa A H 2012 Nanolithography on responsive materials, *J. Nanosci. Lett.* **2**, 23
[5] Wang Xiaohua, Wang Xin, Fernandez R, Ocola L, Yan M and La Rosa A H 2010 Electric field-assisted dip-pen nanolithography on poly(4-vinyl pyridine) films *ACS Appl. Mater. Interface* **2** 2904
[6] Maedler C, Chada S, Cui X, Taylor M, Yan M and La Rosa A H 2008 Creation of nanopatterns by local protonation of P4VP via dip pen nanolithography *J. Appl. Phys.* **104** 014311
[7] Urbakh M, Klafter J, Gourdon D, and Israelachvili J 2004 The nonlinear nature of friction *Nature* **430** 525
[8] Hammer M U, Anderson T H, Chaimovich A, Shell M S and Israelachvili J 2010 The search for the hydrophobic force law *Faraday Discuss* **146** 299
[9] Persson B N J 2000 Sliding Friction (Springer).
[10] Kim S H, Asay D B and Dugger M T 2007 Nanotribology and MEMS *Nano Today* **2** 22
[11] Zhuang Y and Menon A 2005 On the stiction of MEMS materials *Tribol Lett* **19** 111 https://doi.org/10.1007/s11249-005-5088-1
[12] Schenk M, Füting M, Reichelt R 1998 Direct visualization of the dynamic behavior of a water meniscus by scanning electron microscopy *Appl. Phys.* **84** 4880
[13] Weeks B L, Vaughn M W, DeYoreo J J 2005 Direct Imaging of Meniscus Formation in Atomic Force Microscopy Using Environmental Scanning Electron Microscopy *Langmuir* **21**, 8096
[14] La Rosa A H, Nordstrom R, Cui X, McCollum J and Li N 2005 The Ultrasonic/Shear-Force Microscope: Integrating Ultrasonic Sensing into NSOM *Rev. Sci. Instrum.* **76** 093707
[15] Brockman T, Bai J, Fernandez R and La Rosa A H 2018 Assessing probe damage in constant-frequency and frequency-modulation shear-force acoustic near-field microscopy Proc IEEE 13th Nanotechnology Materials and Devices Conference (NMDC), pp. 1-4 doi: 10.1109/NMDC.2018.8605871 https://ieeexplore.ieee.org/document/8605871
[16] Cui X and La Rosa A H 2005 Investigation of the probe-sample interaction in the ultrasonic/shear force microscope: the phononic friction mechanism *Appl. Phys. Lett.* **87** 231907

[17] Wang X, Fernandez R, N. Li, Hung H, Venkataraman A, Nordstrom R, La Rosa A H 2016 Near-field acousto monitoring shear interactions inside a drop of fluid: The role of the zero-slip condition *Phys. Fluids* **28** 052001

[18] Fernandez R, Brockman T, Bai J and La Rosa A H 2018 Confined Fluid Analyzed with Near-field Acoustic Detection *J. Phys.: Conf. Ser.* **1143** 012017 http://iopscience.iop.org/article/10.1088/1742-6596/1143/1/012017/pdf

[19] Supichayanggoon K, Brockman T, La Rosa A H 2023 Acoustic monitoring humidity effects on the formation of water bridges at sharp contacts Journal of Physics: Conference Series 2678 (1), 012010

[20] Yumei Men, Xianren Zhang and Wenchuan Wang. Capillary liquid bridges in atomic force microscopy: Formation, rupture, and hysteresis. *J. Chem. Phys.* ***131*** 184702 (2009); https://doi.org/10.1063/1.3257624

[21] We have considered v = 6310 m/s as the speed of sound in silicon. In water v= 1500 m/s. At 32 kHz the wavelength is l_{water} = 4.77 cm http://www.rfcafe.com/references/general/velocity-sound-media.htm

[22] Hartland S, Hartley R 1976 *Axysymmetric Fluid-liquid interfaces*; Elsevier: Amsterdam, 1976

[23] Griffiths D F, Higham D J 2010. *Numerical Methods for Ordinary Differential Equations: Initial Value Problems* (1st ed.). Springer. https://doi.org/10.1007/978-0-85729-148-6

[24] Karrai K and Tiemann I 2000 Interfacial shear force microscopy *Phys. Rev. B* **62** 13 174-181

Journal of Physics: Conference Series **3052** (2025) 012002 doi:10.1088/1742-6596/3052/1/012002

First principle implementation of the FDTD method for computationally modeling electrodynamics using C++ and Python

Collin A. Bond and Andres H. La Rosa

Department of Physics, Portland State University, PO Box 751 Portland, OR 97207, USA

E-mail: andres@pdx.edu

Abstract. The Finite-Difference Time-Domain (FDTD) method is a brute-force numerical technique for solving a system of differential equations that is perhaps best known from its application in computational electrodynamics as a means of solving Maxwell's equations to model electromagnetic wave interactions with various media. In this paper, we focus on the development of such a model from the ground up by using a partitioned approach that employs both the C++ and Python programming languages; the former handles the numerical computation while the latter addresses data visualization. Following a thorough exploration of the theoretical underpinnings behind this technique, an implementation of the FDTD is constructed directly from first-principles using C++. This dual-language methodology allows us to overcome the limitations of each language while leveraging the strengths of both. After simulating several examples of real-world phenomena, such as thin film interference, the results of our model are compared to the expected analytical results in order to evaluate the model's accuracy. The aim of this work is to highlight the means by which theoretical concepts may be put into practice by providing an introduction to the FDTD method with a sufficient degree of detail as to allow the reader to begin identifying potential avenues for applying this method within their own studies.

1. Introduction

Problems within the study of electromagnetics are notorious for their dependence on the symmetry of the system in question. Consider the classic case of deriving the electric field due to a disk of uniform surface charge density at a point directly above its center. This elementary example becomes a graduate level problem upon simply moving this point off-axis from the disk. With the loss of radial symmetry, the non-vertical components of the field will no longer cancel and the quadrature becomes significantly more difficult. Moreover, this property is far from exclusive to this particular problem—rather, electromagnetics is replete with examples wherein analytical solutions become exceedingly more elusive, or outright unobtainable, after relatively minor alterations to the problem geometry. Of course, one could correctly assert that, rather than presenting a unique phenomena to electromagnetism specifically, this issue is generally true throughout all areas of physics. Although this is true, we assert that the issue remains particularly pernicious to E&M due to the ubiquitous application of electromagnetic theory within today's modern electronic devices. This pervasive influence, coupled with the further imposition of now nanoscale physical dimensions of such devices, underscores the difficulty of obtaining analytical solutions to systems that lack perfect symmetry and highlights the critical importance of developing highly accurate numerical methods to yield solutions under realistic conditions.

The Finite-Difference Time-Domain (FDTD) is a numerical method of solving partial differential equations (PDEs) in the time-domain. Partly due to the widespread commercial success of companies such as Ansys Lumerical and Remcom, FDTD is often ill-conceived as being linked exclusively to electrodynamics. Rather, FDTD is a robust technique which is capable of modeling many different phenomena, including acoustics, fluid dynamics, and even quantum mechanical systems as shown by Perez and Nagel [9]. Despite this versatility, however, we will limit our scope to consider application of the FDTD within the discipline it is best known, electrodynamics.

There is a great deal of existing literature which both explains the FDTD theory and includes details on its implementation—with an excellent example being the work of Schneider in [12]. Such examples, however, may often struggle with walking the line between addressing an audience of experts or novices. Should one assume too great a level of prerequisite knowledge on behalf of the reader, the result is a steep and intimidating learning curve that is difficult to follow for those who are new to the topic. On the other hand, assuming too little may result in an audience confused by seeing new notation and conventions for a topic they are already somewhat familiar while being left ultimately unsatisfied due to a lack of depth that is sufficient for them to have connected all the dots. Hence, despite the quality of the existing publications, these resources may prove ill-suited to the specific role as a stand-alone introduction to the topic. It is our opinion that such an introduction must:

1. Demonstrate the logic behind how the central difference approximation is used as a vehicle to obtain a discretized form of a parabolic PDE—allowing it to be numerically solved on a finite grid of discrete time and space.
2. Explain the mathematical implications of using the central difference and discuss what considerations should be taken to address the issues they present for the FDTD.
3. Make the explicit connection between this theory and how it can be implemented programmatically. Demonstrate that the model is consistent with our analytical methods and may also be used to simulate more complex phenomenon.
4. Leave the reader with the following: (A) a sufficient understanding of the theory to begin seeing the FDTD as a general method that can be used to solve other parabolic PDEs. (B) to see how it can be applied as a general method to solve other parabolic PDEs as well as enough familiarity with the code to begin building a simple model of their own.

As the tools offered by Ansys Lumerical and others become increasingly capable, complex, and prevalent within the scientific community, the need for a solid foundation on the inner-workings is made only *more* important. While the utility afforded by the invention of the calculator cannot be understated, none would argue against the necessity for one to be capable of performing multiplication by-hand. Without such understanding, however, these tools threaten to become a *black box*—data simply goes in, data simply comes out. All questions about the inner-workings of the box are silenced by the sheer scale and complexity of the device and, similarly, all answers are obfuscated in the same manner. The use of any device in such a manner stands in direct contradiction to the scientific method. As such, we believe that the details required for a first-principle implementation of the FDTD are not only non-trivial—but rather the cornerstone from which we should begin our study of the topic and a necessary precondition to using the increasingly available tools of our day.

This paper presents the first installment of a series in which we address new methodologies which allow us to capitalize on the virtues of runtime performance afforded by C/C++ in addition to the ease-of-use and widespread portability of Python for our FDTD implementation.

2. Theoretical Background

The crux of the FDTD algorithm lies in the central difference numerical approximation of the derivative—which, as we will see, indirectly computes the derivative at a point x_0 by sampling the adjacent points $x_0 \pm \delta$. In the world of continuous time and space, this detail is fairly innocuous; for discrete values, however, this is integral to the FDTD algorithm. Our ultimate goal is to express the partial derivatives in Maxwell's equations with the central difference, thereby obtaining a representation of these equations on a discrete grid of time and space. The FDTD functions by clever application of a *leapfrog* technique, which exploits this discrete grid.

2.1. Central Difference Approximation

Consider the Taylor expansion of an arbitrary, differentiable function, $f(x)$, about the point x_0 with an offset of $\delta/2$,

$$f(x_0+\delta/2) = f(x_0) + \frac{\delta}{2}f'(x_0) + \left(\frac{\delta}{2}\right)^2 f''(x_0) + \frac{1}{2!}\left(\frac{\delta}{2}\right)^3 f'''(x_0) + \mathcal{O}(\delta^4)\,, \tag{1}$$

where $\mathcal{O}(\delta^4)$ defines the higher-order terms in the expansion. By subtracting the Taylor expansion for $f(x_0-\delta/2)$, many of these terms cancel and we are left with,

$$f(x_0+\delta/2) - f(x_0-\delta/2) = \delta f'(x_0) + \left(\frac{2}{3!}\right)\left(\frac{\delta}{2}\right)^3 f'''(x_0) + \cdots .$$

Dividing through by δ,

$$\frac{f(x_0+\delta/2) - f(x_0-\delta/2)}{\delta} = f'(x_0) + \frac{\delta^2}{3!2^2}f'''(x_0) + \cdots + \mathcal{O}(\delta^2)\,.$$

We see that the left-hand side (LHS) of the equation above mimics the first principal definition of the derivative, $\frac{df}{dx}|_{x_0}$. Since higher-order terms have smaller influence in the expansion, neglecting second order terms yields an approximation for the derivative as,

$$\left.\frac{df}{dx}\right|_{x_0} \approx \frac{f(x_0+{}^{\delta}/_{2}) - f(x_0-{}^{\delta}/_{2})}{\delta}\,. \tag{2}$$

Eq. (2) above is known as the central difference numerical approximation of the derivative. Despite computing the derivative at x_0, the function itself is never sampled at x_0 directly. In essence, the FDTD algorithm functions by discretizing time and space in such a manner as to exploit this very property.

2.2. Maxwell's Curl Equations

We begin with the Faraday and Amperè-Maxwell laws of Maxwell's equations in MKS units as,

$$\boldsymbol{\nabla}\times\mathbf{E} = -\frac{\partial\mathbf{B}}{\partial t} = -\mu\frac{\partial\mathbf{H}}{\partial t}\,, \tag{3}$$

$$\boldsymbol{\nabla}\times\mathbf{B} = \mu_0\mathbf{J} + \mu_0\epsilon_0\frac{\partial\mathbf{E}}{\partial t}\,, \tag{4}$$

where $\mathbf{E}$ is the electric field, $\mathbf{B}$ is the magnetic field, $\mathbf{J}$ is the current density, μ_0 is the permeability of free space, and ϵ_0 is the permittivity of free space. We will implement our FDTD model for a single spatial dimension, given by the x-axis, where E_z is only non-zero component of the electric field. Moreover, we assume a source-free region in which $\mathbf{J}=0$ and $\rho_f=0$. Under these assumptions, the Faraday and Amperè-Maxwell laws reduce to the following,

$$-\mu\frac{\partial\mathbf{H}}{\partial t} = -\frac{\partial E_z}{\partial x}\hat{\mathbf{y}}\,, \tag{5}$$

$$\epsilon\frac{\partial E_z}{\partial t} = \frac{\partial H_y}{\partial x}\hat{\mathbf{z}}\,. \tag{6}$$

From Eq. (5), H_y is the only non-zero component of the time-varying magnetic field, $\mathbf{H}$, since the RHS has only a y-component. Furthermore, note that each equation neglects static terms of $\mathbf{H}$. From the two expressions above, we obtain the following pair of scalar relations,

$$\frac{\partial H_y}{\partial t} = \frac{1}{\mu}\frac{\partial E_z}{\partial x}\,, \tag{7}$$

$$\frac{\partial E_z}{\partial t} = \frac{1}{\epsilon}\frac{\partial H_y}{\partial x} \tag{8}$$

Eq. (7) relates the time derivative of $\mathbf{H}$ to the spatial derivative of $\mathbf{E}$ while Eq. (8) relates the time derivative of $\mathbf{E}$ to the spatial derivative of $\mathbf{H}$. In the interest of brevity, we may simply refer to these equations as our *scalar relations.* Taken together, we begin to see the origins for the *leapfrog* technique that we will later address—wherein one quantity is used to obtain the other, and vice-versa again.

2.3. Discretization

Following the conventions of Schneider in [12], discretized quantities in Maxwell's equations are denoted,

$$E_z(x,t) = E_z(m\Delta_x,\, q\Delta_t) \stackrel{\text{def}}{=} E_z^q[m]\,. \tag{9}$$

Here, time and space are discretized as $t=q\Delta_t$ and $x=m\Delta_x$, where $q,m\in\mathbb{N}$ represent the temporal and spatial indices as integer multiples of the fundamental, minimum stepsizes, Δ_t and Δ_x. Placing the values of q in the superscript emphasizes that each computation is performed at a fixed value of time. Brackets are used in place of the traditional parenthetical function arguments to better distinguish the discrete quantity from its continuous counterpart as well as to mimic the C-style programming syntax wherein, for example, the subscript operator is used as "`values[m]`" to index the mth value in the array named "`values`". Using this notation to express $\partial E_z/\partial x$ in Eq. (7) as a finite difference gives us,

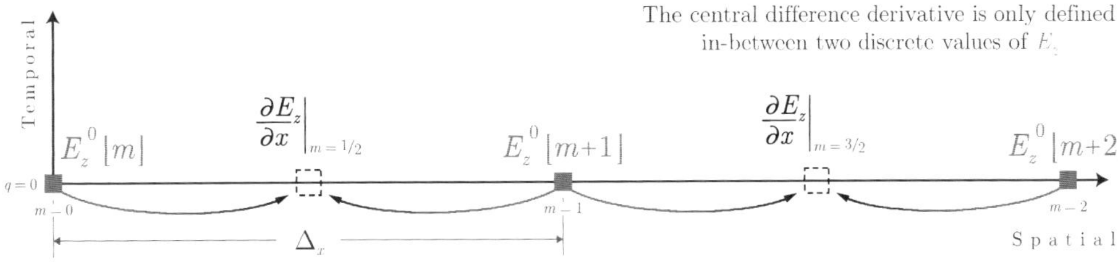

Figure 1. Staggered spatial position of $\partial E_z/\partial x$ relative to discrete values of E_z due to the central difference derivative. Discrete values of E_z are shown as ■; $\partial E_z/\partial x$ is shown as dashed-□.

$$\frac{\partial H_y}{\partial t} = \frac{1}{\mu}\frac{\partial E_z}{\partial x}\bigg|_{\left((m+\frac{1}{2})\Delta_x,\ q\Delta_t\right)} = \frac{1}{\mu}\left(\frac{E_z^q[m+1]-E_z^q[m]}{\Delta_x}\right). \tag{10}$$

It is critical to note that the expression above represents the value of the derivative at the spatial index of $m+1/2$—that is, *in-between* $E_z^q[m]$ and $E_z^q[m+1]$, as illustrated in Fig. 1 below.
Hence, on a discrete grid, the spatial derivative of E_z *cannot* exist at the same points as E_z itself—rather, the derivative is spaced in-between two consecutive values of E_z. This result is given as a direct consequence of the central difference's property of computing the derivative at x_0 by sampling the *adjacent values* of $x_0-\delta$ and $x_0+\delta$, as we described earlier.

2.3.1. Playing Hopscotch Looking back to Eq. (7), our goal was to indirectly compute $\partial H_y/\partial t$ by, instead, computing $\partial E_z/\partial x$. As we just demonstrated, however, $\partial E_z/\partial x$ is spatially offset by a half-integer index relative to E_z. Given that we compute H_y from the offset-derivative of E_z, this implies that H_y *must also be offset by the same amount.* In other words, **E** and **H** do not exist at the points in space. Rather—each field is spatially offset from the other by a factor of $\Delta_x/2$.

By the same logic, Eq. (8) has us compute $\partial E_z/\partial t$ via $\partial H_y/\partial x$. Since H_y is spatially staggered as $H_y[m+1/2]$, the finite difference for Eq. (8) becomes,

$$\frac{\partial E_z}{\partial t} = \frac{1}{\epsilon}\frac{\partial H_y}{\partial x}\bigg|_{\left(m\Delta_x,\ (q+\frac{1}{2})\Delta_t\right)} = \frac{1}{\epsilon}\left(\frac{H_z^{q+\frac{1}{2}}[m+1/2]+H_z^{q+\frac{1}{2}}[m-1/2]}{\Delta_x}\right). \tag{11}$$

Therefore, we see that **H** and **E** are staggered in both space and time by $\Delta_x/2$ and $\Delta_t/2$, respectively. This property results in the emergence of the staggered grid, which was first published in 1966 by Kane Yee in [15], which we will discuss in section 2.4.

2.3.2. The Update Equations By expressing each derivative in the scalar relations from Eq. (7) and Eq. (8) as a finite difference in the manner shown above, we are able to algebraically manipulate the equation to solve for the value of these fields at one time-step into the future. The equations which result from this computation are derived by Schneider [12] as,

$$H_y^{q+\frac{1}{2}}[m+1/2] = H_y^{q-\frac{1}{2}}[m+1/2] + \frac{\Delta_t}{\mu\Delta_x}\left(E_z^q[m+1]-E_z^q[m]\right), \tag{12}$$

and,

$$E_z^{q+1}[m] = E_z^q[m] + \frac{\Delta_t}{\epsilon\Delta_x}\left(H_y^{q+\frac{1}{2}}[m+1/2]-H_y^{q+\frac{1}{2}}[m-1/2]\right), \tag{13}$$

for the electric and magnetic fields, respectively. Although the notation may appear intimidating, each equation is fundamentally quite simple—in essence: *the value of the field at one time-step into the future is given by the present value of the field plus the amount by which the field changed in a single time-step,* Δ_t. Hence, each equation is simply stating,

$$H_y^{q+\frac{1}{2}+1} = H_y^{q+\frac{1}{2}} + \frac{\partial H_y}{\partial t}\bigg|_{\Delta_t}, \qquad \text{and} \qquad E_z^{q+1} = E_z^q + \frac{\partial E_z}{\partial x}\bigg|_{\Delta_t}.$$

Journal of Physics: Conference Series **3052** (2025) 012002 doi:10.1088/1742-6596/3052/1/012002

The confusion that arises is largely due to the obfuscation introduced by having terms of both **H** and **E** in each expression. These terms, however, are simply a result of our *leapfrog* technique of using our scalar relations along with a finite difference to obtain the the time-derivative of one field via the spatial-derivative of the other. As such, these equations are commonly referred to as the *update equations* due to their role in using the current values of **E** to compute the future value of **H**, and vice-versa—thereby *updating* each quantity in the simulation.

2.4. Kane Yee's Algorithm

The reasons that spatial and temporal discretization are required for the Yee algorithm are given by the following:

(1) Assume the initial values of E_z and H_y are known at $t = 0$. We will denote these as E_z^0 and H_y^0.

(2) The algorithm begins by computing the future values $H_y^{q+\frac{1}{2}}$ from the initial conditions.
  2.1 First, $\partial E_z/\partial x$ is computed with the central difference, as illustrated in Fig. 1.
  2.2 Then, $\partial H_y/\partial t$ is computed from $\partial E_z/\partial x$ with our scalar relation in Eq. (7).
  2.3 Finally, recognizing that that $\partial H_y/\partial t$ provides a measure for the amount that H_y changes over the time Δ_t, we obtain the future value by incrementing the initial value by the derivative as, $H_y^{q+\frac{1}{2}} = H_y^{q-\frac{1}{2}} + \partial H_y/\partial t$.

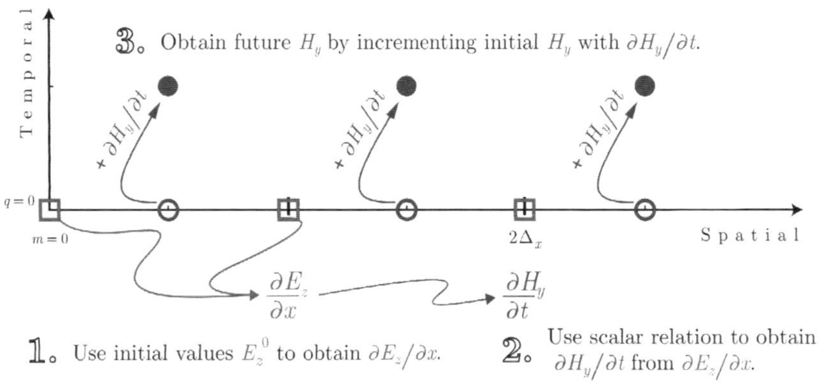

Figure 2. First step of the leapfrog technique—initial values of E_z are used to compute future values of H_y. □ and ○ are the initial $E_z^0[m]$, $H_y^0[m]$ simulation values, and ○ are discrete values of $H_y[m]$.

(3) Next, the future values of E_z^{q+1} are computed from the $H_y^{q+\frac{1}{2}}$ values from the last step, as $E_z^{q+1} = E_z^q + \partial E_z/\partial t$.

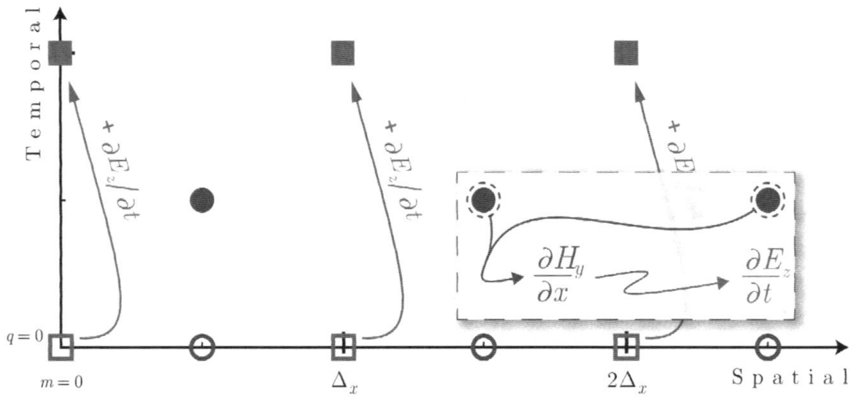

Figure 3. Second step of the leapfrog technique. Current values of H_y are used to compute future values of E_z.

(4) The value of q is incremented and steps (2) through (4) are repeated until the desired number of timesteps have elapsed. That is, we continue applying the leapfrog method.

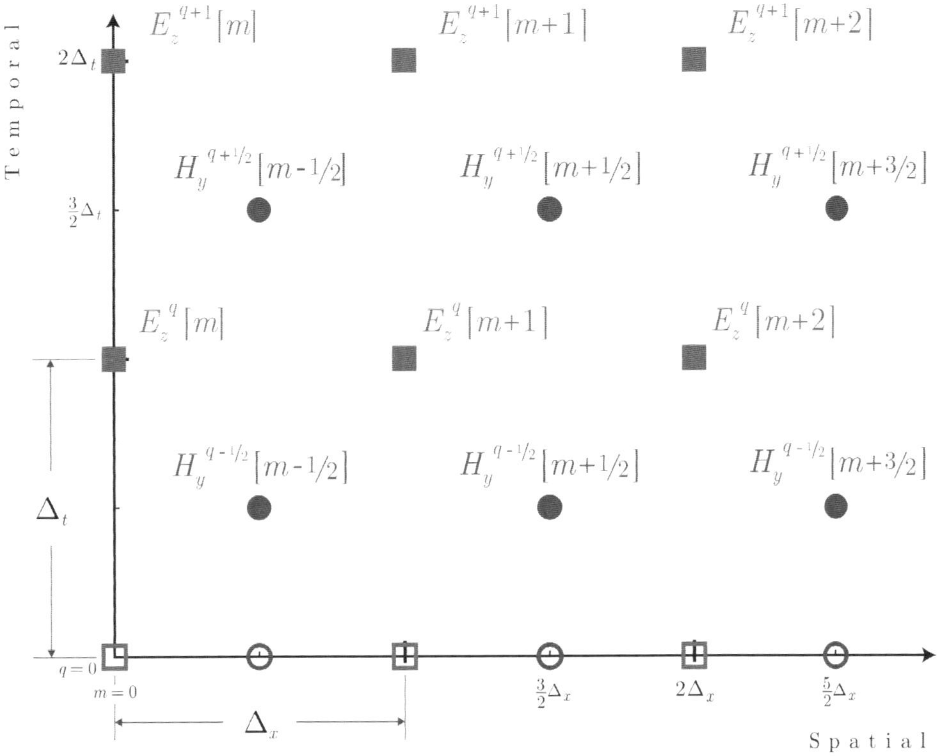

Figure 4. Complete illustration of Yee's staggered grid with the same conventions as Fig. 2.

2.5. Courant-Friedrichs-Lewy (CFL) Conditions

The staggered/leapfrog nature of the Yee grid carries several critical implications for our selection of the physical of the simulated components and the simulation duration. For other numerical methods, it is often the case that further reducing the stepsize yields increased numerical accuracy that follows the law of diminishing returns. With *two* stepsizes, Δ_x and Δ_t which are related by the leapfrog operation, however, this result is not as simple.

In our update equations for **H** and **E**, we see a term of $\Delta_t/\mu\Delta_t$ and $\Delta_t/\epsilon\Delta_t$, respectively. Coining a term for these expressions as the *update coefficients*, Schneider goes on to state that "[it] is often convenient to represent [...]" these quantities by "[...] the ratio of how far energy can propagate in a single temporal step [...]" to the minimum separation distance between each "[...] spatial step" [12]. Schneider carries on to make the assumption that the maximum distance that energy may travel within a time Δ_t is Δ_x. Given that the maximum velocity for an EM wave is c, this assumption yields the following expression for the Courant number[1],

$$S_c = \frac{c\Delta_t}{\Delta_x} = 1\,. \tag{14}$$

Using the Courant number above, we are able to express our update coefficients in terms of the for our update coefficients as,

$$\frac{1}{\epsilon}\frac{\Delta_t}{\Delta_x} = \frac{1}{\epsilon_r\epsilon_0}\frac{\sqrt{\epsilon_0\mu_0}}{\sqrt{\epsilon_0\mu_0}}\frac{\Delta_t}{\Delta_x} = \frac{\eta_0}{\epsilon_r}S_c\,, \qquad \frac{1}{\mu}\frac{\Delta_t}{\Delta_x} = \frac{1}{\mu_r\mu_0}\frac{\sqrt{\epsilon_0\mu_0}}{\sqrt{\epsilon_0\mu_0}}\frac{\Delta_t}{\Delta_x} = \frac{1}{\mu_r\eta_0}S_c\,. \tag{15}$$

[1] Notice that this expression only is valid in 1-D. Increasing the dimensionality would dictate that traveling a net distance of Δ_x in a time Δ_t would, instead, be $\sqrt{\Delta_x^2 + \Delta_y^2 + \Delta_z^2}$.

By reducing our discretized expressions to forms in which no terms of Δ_x or Δ_t are present, we are able to observe the FDTD stability behavior without explicit need for specifying spatial or temporal dimensions. This approach is especially useful during the early stages of development when it is difficult to identify the presence of errors in the code.

To summarize, it is advantageous to obtain expressions which are agnostic of the specific numerical value of a stepsize. Instead, we abstract the simulation to function in terms of the stepsize *multiple*, m or q, rather than the explicit numerical value of the product $m \cdot \Delta_x$ or $q \cdot \Delta_t$.

This approach, however, has the clear disadvantage of being entirely decoupled from any physical dimension for the situation, as the specific value of Δ_x and Δ_t are not computed for the simulation. Though this may seem undesirable for a technique of simulating real-world phenomena, the reason for this is simple. Selecting specific values for each step-size is an art unto itself. Although there are general guidelines for success when computing these specific values, such as applying the Nyquist–Shannon sampling theorem, whether or not the specific values will produce Courant stability in the simulation is subject to a number of factors—including numerical error and the minutia of the specific code itself. Even minor alterations to these values can turn a functional simulation into a numerical mess that diverges over just a few time-steps. This volatility makes it exceedingly difficult to troubleshoot the model; worse yet, this issue may serve to conceal more fundamental errors for a lucky set of stepsize values—creating a house of cards which appears to function correctly yet is destined to collapse upon the slightest change to a certain parameter later on. To be explicit, our rationale is this: it is best practice to achieve an accurate model using abstract terms for each stepsize and *then* graduate to use of specific numerical values with a simulation we already know to be working, than it is to begin with numerical values from the beginning.

3. Methods and Implementation

With the discretized forms of Maxwell's equations we derived above, we are now able to construct our first programmatic implementations of the FDTD algorithm. The operations of our model can be broadly classified under one of two categories: one is that of numerical computation and the other is that of data visualization. Each responsibility is handled by its own, distinct program: one written in C++ and the other in Python, respectively. That is, the C++ program implements all numerical methods, performs all computations for the FDTD algorithm, and saves the numerical results to an external data file in binary format. Subsequently, the Python program loads this data and constructs any requested visualizations, such as plots or video animations.

3.1. A Minimal Working Example (MWE) Program

The implementation details of our model are much too lengthy to present here in their entirety. Moreover, a great deal of code pertains to the logistical tasks rather than the core computation of the FDTD. To avoid such obfuscation by focusing on these aspects, the excerpts of code are intended to convey the most fundamental operations needed to implement the FDTD in their most general form. In the interest of brevity, the code in Listing 1 below is treated as a preamble that we assume to be included with each subsequent excerpt we address.

```
// 1. Main Model Parameters.
constexpr std::size_t   NT     = 200ULL;// Total length of simulation (time-steps)
constexpr std::size_t   NX     = 200ULL;// Number of nodes on x-axis
constexpr double        src_x  = 100ULL;// Position of the source
constexpr double        eta_0  = 377.0f;// Characteristic impedance of free space

// 2. Typename Aliases.
using complex_t                = std::complex<double>;
using re_array                 = std::array<double, NX>;
using im_array                 = std::array<complex_t, NX>;
using frame_array              = std::array<re_array, NT>;

// 3. FDTD Variables.
re_array              Hy,        Ez;//  Field values at a single timestep
frame_array           Hy_T,      Ez_T;// Field values at every timestep
im_array              eps_r,     mu_r;// Permittivity and permeability values
```

Listing 1. Preamble declarations to be included with each subsequent excerpt of code.

With the preamble above, we move to our first FDTD program in Listing 2 below:

```
1  //  Main FDTD Loop...
2  for (std::size_t q = 0ULL; q < NT; ++q) {
3      for (m = 0ULL; m < NX - 1; ++m)                        // 1. Update H-Field.
4          Hy[m]       = Hy[m] + (Ez[m+1] - Ez[m]) / eta_0;
5
6      for (m = 1ULL; m < NX; ++m)                            // 2. Update E-Field.
7          Ez[m]       = Ez[m] + (Hy[m] - Hy[m-1]) * eta_0;
8
9      Ez[src_x] = std::exp( -std::pow( (q - src_d)/src_w, 2) );  // 3. Hardwired Source.
10
11     for (m = 0ULL; m < N_x; ++m) {       // 4. Store the current Frame Data.
12         Hy_T[q][m]  = Ez[m]
13         Ez_T[q][m]  = Ez[m]
14     }
15 }
```

Listing 2. MWE program / main loop of the FDTD procedure.

Notice that, due to the staggered nature of the Yee grid—with a finite number of spatial nodes, N_X, the left endpoint must be a different quantity than the right. That is—if the 0th-node measures E_z, then the N_X^{th}-node must measure H_y, or vice-versa.

The crux of this implementation lies in the indices used for the H_y and E_z update loops on lines 3 and 6. Despite the fact that each field varies with both time and space, the program above uses only one-dimensional arrays to store H_y and E_z. This design leverages the fact that the residual values from the previous time-step will remain within the `Hy` and `Ez` arrays until each value is subsequently overwritten upon the next iteration. That is—the statement `Hy[m] = Hy[m] + ...` is equivalent to computing the future field value as `Hy[q + 1][m] = Hy[q][m] + ...` as the value computed by the previous `q`-iteration will survive inside the `Hy` array until it is overwritten in the next `m`-iteration. This approach, however, requires us to save each frame to a secondary data structure with a longer-lifetime to avoid losing the previous values when they are overwritten. This operation is performed in `step 4` above.

3.1.1. Simulating Materials and Inhomogeneities By using the alternative form for our update coefficients in Eq. (15), we introduce the values of ϵ_r and μ_r into our update equations. Originally, the simulation was constructed to model a uniform medium of vacuum. To model inhomogeneities, we create two array of N_x-number of elements—one for ϵ_r and one for μ_r. In other words, each x-node on the Yee Grid contains a corresponding permittivity and permeability value.

3.2. Consistency With Analytical Methods

Perhaps the most challenging aspect of implementing the FDTD is assessing the extent to which the model aligns with reality. That is, it can be particularly easy to produce compelling data that, upon first glance, appears to be quite convincing upon first glance, but, upon close and careful inspection, is entirely wrong. As such, developing a suite of tests and comparisons which compare the simulation behavior against that of known, analytical results and real-world phenomena to ensure that the model results are consistent to real-world known phenomena, both qualitatively and quantitatively, is imperative. Given that the entire purpose of a numerical method is to yield solutions to examples in which analytical methods are not available, it is challenging to adhere to this standard. However, in the simpler case, developing a small set of comparisons with known analytical behaviors allows us to benchmark the performance and accuracy of the model—giving us a reference point from which we may assess its accuracy when modeling a more complex system.

3.2.1. Model Assumptions In consideration of our scope—several assumptions are made to simplify the electrodynamic systems we will simulate. First, the only source in our region is given by the incident wave $\rho_f = \mathbf{J}_f = 0$. Second, all media are linear, homogeneous, and non-magnetic. Third, all waves are monochromatic plane-waves and the effects of dispersion are neglected.

Although a greater number of real-world phenomenon are available to simulate with a two or three dimensional model, the topics of thin film interference, transmission and absorption, and frequency domain behaviors offer a great deal of utility within a 1-dimensional context.

3.2.2. Thin Film Interference Consider two linear media of refractive indices n_1 and n_2 that are joined at an interface. When a wave travels from n_1 to n_2, part of the wave is transmitted *through* the interface and continues to propagates within n_2 while the other reflects *off* and remains inside n_1. If $n_2 > n_1$, the amplitude of the reflected wave is inverted from that of the incident. Inside n_1, however, this reflection interferes with the incident wave, introducing distortion and other artifacts as a result. Now, suppose that n_2 has a finite thickness of $d \mid d \sim \lambda_0$, after which the wave encounters a second interface transitioning to a *third* medium, n_3 with an index of $n_3 = n_1$. Upon exiting n_2, another instance of reflection/transmission occurs—since $n_2 < n_3$, however, the second reflection is *not* inverted. In addition to interfering with the incident wave, the reflections at each interface, in turn, interfere with each other—creating a complex array of many wave interactions. However, the aggregate effect of these interactions can be represented as either one of two cases: constructive interference, wherein the amplitude of the *transmitted* wave is maximum, or destructive, where the amplitude of the *reflective* wave is maximum.

$$2nd = N\lambda_0\,, \tag{16}$$ $$2nd = \big(N + 1/2\big)\lambda_0\,, \tag{17}$$

respectively, where $n = \sqrt{\epsilon_r \mu_r}$ is the refractive index of the film, λ_0 is the wavelength of incident light in a vacuum, and $N \in \mathbb{Z}^+$. In one-dimension, of course, we may only consider the case of oblique or normal incidence.

The emergence of a maximum or minimum in transmission is not explicitly described by the equations solved in the FDTD. Rather, this behavior is an emergent property of the continued application of Maxwell's equations to compute each individual wave interaction in the system. As such, thin film interference presents an excellent test for evaluating the accuracy of our model as any discrepancies we observe between our results and the analytical model are likely to reveal the presence of any errors that may be subtly hidden within our code.

Since λ and d are measurements of a physical distance, we may express them as some integer multiple of Δ_x. To model this phenomenon, we implement a time harmonic source with wavelength $\lambda_0 = 40\Delta_x$ and a unity amplitude which emits from the node $m = 2$. From Eq. (16) and Eq. (17)—with a lossless medium of relative permittivity $\epsilon_r = 4$, the widths of $d_1 = 10N\Delta_x = \{10,\, 20,\, 30,\, 40,\, \cdots\}$ will exhibit constructive interference while widths of $d_2 = 5N\Delta_x = \{5,\, 15,\, 25,\, 35,\, \cdots\}$ will exhibit destructive interference. For simplicity, We select the widths of $d_1 = 30\Delta_x$ and $d_2 = 25\Delta_x$ position all films at the $m = 52$ node (50 nodes ahead of the source).

3.3. Program Design

This decision to use two independent programs provides several key advantages for both development of the model and analysis of our results. The programming required to implement the FDTD procedure is very subtle in several aspects.

3.3.1. Numerical Stability and Performance Considerations First is the aspect of numerical instability and error within the method. A major fundamental issue that plagues the FDTD is that of numerical instability and truncation error. the most difficult issue upon first implementation of the FDTD is effectively managing the CFL convergence conditions, as previously discussed. Even small numerical errors in the simulation present major issues to the simulation accuracy as such errors compound with each iteration of the algorithm. This issue manifests as a persistent cumulative error within the accuracy of our model. The inherent truncation issues that arise from finite precision floating-point values in a digital environment coupled with the numerical instability of the FDTD algorithm mix together as well as oil and water. C++ provides primitive data types for higher precision floating-point values, such as the `long double`. Although standard Python packages offer extended precision floats as well, consistent support of these types by auxiliary packages, such as the `struct` package which is used in our implementation to interface with the binary files produced in C++, is somewhat limited. Furthermore, the direct memory access afforded by low-level languages like C/C++ allows for the implementation of an entirely custom templated floating-point class to implement decimal values of arbitrary precision. Not only would this be difficult to achieve in Python, the additional costs of operating with larger memory-sizes for each value would only serve to exacerbate the existing issue of runtime performance in a Python implementation.

3.3.2. Debugging and Preventing Error Second is the property wherein even the most subtle changes to the code can frequently cause fundamental changes to the simulation. For example, when implementing a source, the difference between a simple `=` or a `+=` statement determines whether energy is able to transmit across either side of the source or not. Similarly, the addition or removal of even a single index within the update equation loop can implement a perfect magnetic conductor (PMC) boundary condition in the simulation. Examples like these, wherein minor changes to the code cause fundamental changes to the simulation, are present while implementing the FDTD for a model of nearly any sophistication. Although a major reason that Python has become so pervasive within the programming community is largely due to its incredible ease-of-use, this interface contingent upon the abstraction of many lower-level programming details out of the user's hands. While this feature is undoubtedly wonderful within many disciplines, the same is not true for the subtle nature of FDTD—rather, this very abstraction serves only to conceal the subtle errors in the FDTD, creating an obstacle which greatly hampers one's introduction to the method. Without a compiler to catch some of the more subtle errors here, there are many cases in which Python will simply execute the program without any notice to this undefined behavior. Simply put, the low-level workflow of C++ is more conducive to catching such errors ahead of time rather than allowing them to remain hidden within the model until their effects are noticed elsewhere.

With this said, however, we wish to be explicit about the following: by no means do we assert that there exists a fundamental limitation which prevents Python from achieving the same results as C++ here. On principle, there is no reason as to why this model could not have been implemented entirely with Python alone. The problem we speak to is one of pragmatism: upon one's introduction to the FDTD, the workflow of C++ is simply more conducive to achieving success than that of Python. Alternatively, for those already familiar, the benefits of allocating the vast and expansive number of computations to a low-level language only accrue as the simulation complexity increases—allowing our model to scale more elegantly as development continues.

4. Data Reduction and Results

4.1. Results from our MWE

Fig. 5 displays a *waterfall plot* that is generated from our MWE program in Listing 2. These plots are an effective means of conveying the time evolution of the system when we are unable to display video animation from our simulation[2]. Each plot represents E_z across the entire spatial dimension of the simulation. Instead of plotting the amplitude on the vertical axis, however, we plot the time-step value corresponding to each plot of the field. To account for the amplitude information we lose from this decision, each plot is normalized with respect to its maximum value to allow the plot to convey shifts in the relative change in amplitude across each time iteration of the simulation. Hence, this plot sacrifices a degree of detail regarding the absolute value of amplitude in order to illustrate the time behavior of the field across the the simulation duration.

[2] The video animations produced directly by our model are available for access at the following URLs: `example_1`, `example_2`, and `example_3`.

Journal of Physics: Conference Series **3052** (2025) 012002 doi:10.1088/1742-6596/3052/1/012002

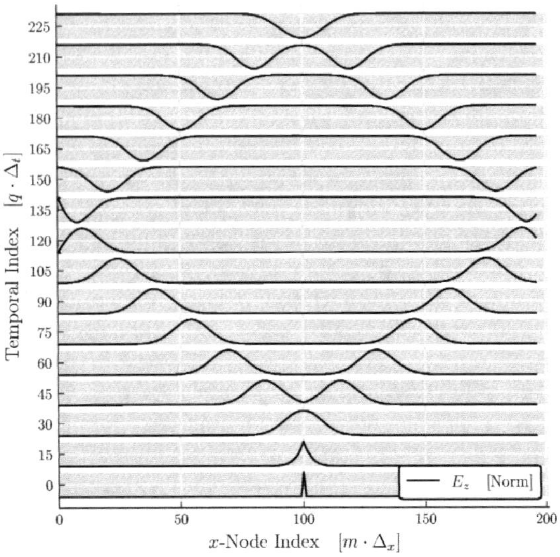

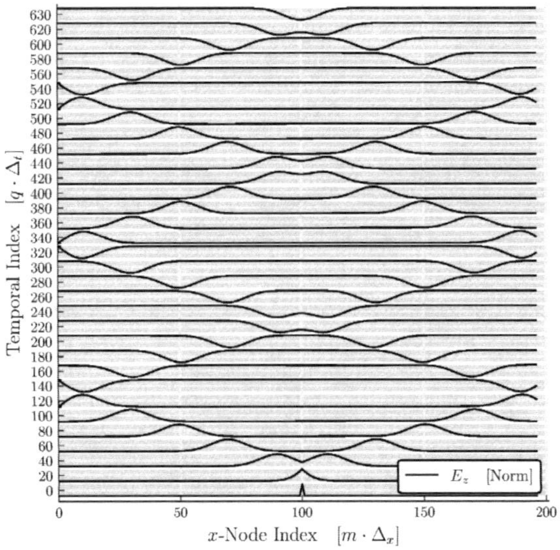

Figure 5. Plot of E_z at each increment of 15 time-steps for the MWE program in Listing 2.

Figure 6. Additional time-steps from Fig. 5 [video is available *here*].

From Fig. 5 above, the Gaussian pulse source is emitted from the central spatial node, $m = 100$, and gradually increases in amplitude until time-step $q = 30$. After this, the Gaussian splits into two pulses that travel along both directions of the x-axis. Upon reaching the endpoints of the plot, the pulse is reflected and travels back towards the center with an inverted amplitude. After each plot meets again at the $m = 100$ node, they each proceed outward, again, toward the endpoints. When the inverted pulses reach the boundaries for the second time, their amplitude is, once again, inverted—yielding back the original condition of the simulation. That is, this simple model depicts a Gaussian pulse which resonates in a lossless medium. The cyclic nature of the system is depicted more clearly in Fig. 6.

4.2. Constructive Interference

Fig. 7 depicts a plot of E_z for the case of $d = 30\Delta_x$ while the adjacent plot in Fig. 8 depicts H_y under the same conditions. Although our scope pertains to non-magnetic media, this plot is simply included to emphasize the utility of being able to plot both quantities in the FDTD method; hence, further analysis of H_y will be excluded from our discussion. The plot shown in – – – is included to illustrate the wave propagation over one-half its period.

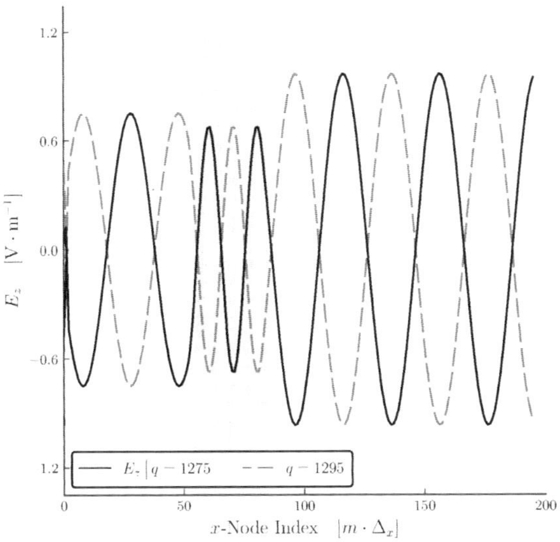

Figure 7. Two consecutive plots of E_z for constructive interference with $d_1 = 30\Delta_x$.

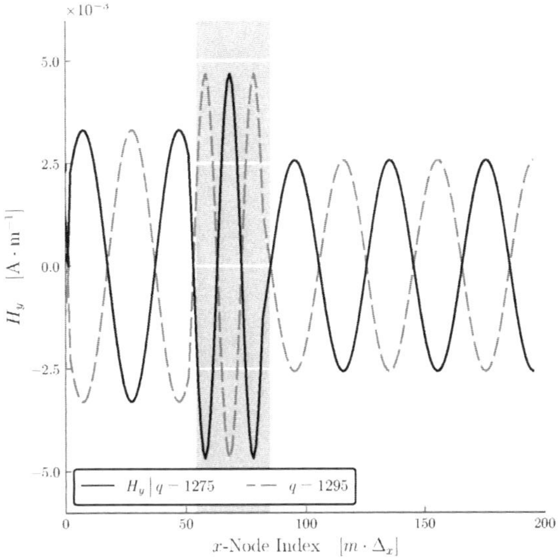

Figure 8. Plot of H_y from Fig. 7 [video *here*].

At first glance, the wave appears to have greater amplitude on the right side of Fig. 7 than the left—which would align with our analytical model of constructive interference. However, this can be misleading as these plots are the product of taking a screenshot of the field at these specific time-steps. To correctly examine the relative amplitude of each wave, we must probe the value of each field at specific nodes at either side of the film, as depicted Fig. 11 below.

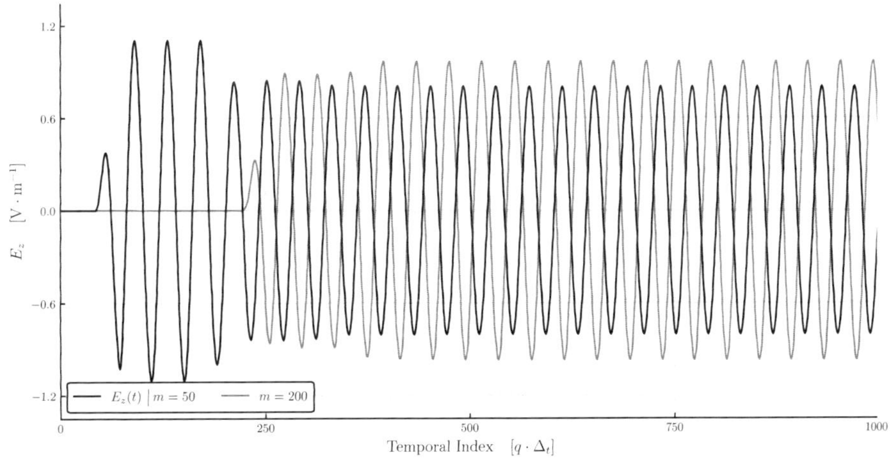

Figure 9. Amplitude of the E_z field at the nodes $m = 50$ [before the thin film] and $m = 200$ [after the film] $d = 30\Delta_x$.

In Fig. 9 above, E_z is measured or *probed* at a fixed point on the x-axis and plotted as a function of time across the simulation duration. One may think of these plots as a buoy in the ocean, rising and falling with the changing height of the ocean waves. Notice the initial delay of 50 and 200 time-steps in the black and red plots, respectively, as this corresponds to the arrival time to reach these nodes after the wave was emitted from node $m = 2$. In this figure, we see that, ultimately, the amplitude of the transmitted wave exceeds that of the incident wave, confirming our predictions for constructive interference due to the width of $d = 30\Delta_x$. Moreover, notice that a transient wave of greater amplitude escapes into the third. This transient is an artifact of the time required for the system to reach a steady state wherein the wave makes an initial pass through the material and each reflection begins to interfere as we expect.

4.3. Destructive Interference

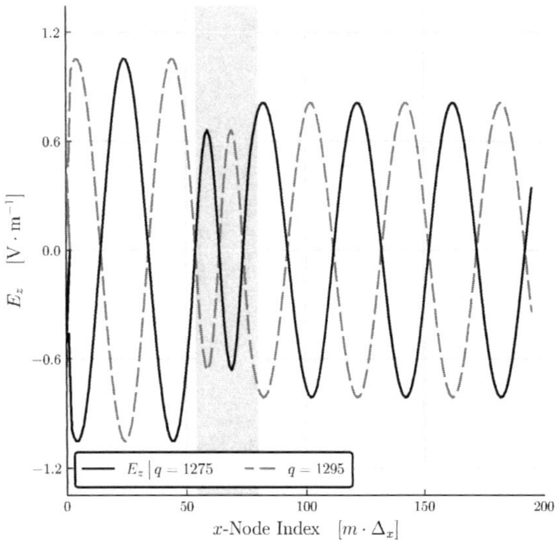

Figure 10. Two consecutive plots of E_z for destructive interference with $d_2 = 25\Delta_x$ [video *here*].

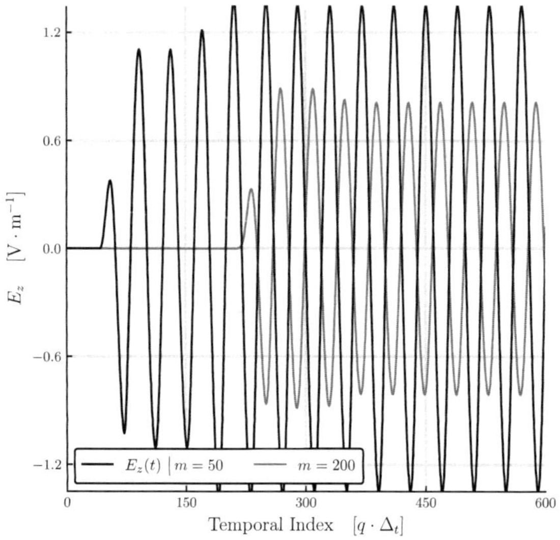

Figure 11. Amplitude of the E_z field at the nodes $m = 50$ [before the thin film] and $m = 200$ [after the film] for $d = 25\Delta_x$.

5. Conclusions

There is a great deal of room for future study regarding the work we have presented here. Without even the introduction of additional complexities to our model, there have already been a plethora of topics we have had to exclude from our discussion in consideration of our scope. Such topics include the frequency domain behavior of our simulation and the effects of dispersion, obtaining the transmission and reflection coefficients, simulating additional material properties such as conductivity, dissipativity, or even non-linear behaviors—just to name a few. When we consider elevating the model to simulate two or even three spatial dimensions, each of these topics only demands that much more consideration and attention

Although our scope has been limited to the specific area of electrodynamics, this is by no means the horizon of the work we have presented. Rather, our intent has been to present the FDTD rationale in a manner which provides a sufficient degree of detail for the reader to begin identifying the means by which a problem within any subject may be formulated in a manner conducive to the application of this technique. Problems within countless areas of physics are described by PDEs which relate the derivative of a certain quantity with respect to a certain variable to the derivative of a second quantity with respect to another. Wherever such relations are found, it is likely that there exists a method of applying Yee's staggered grid in a manner that yields numerical solutions via the FDTD.

Acknowledgments

The work presented throughout this paper would not have been possible were it not for the support, guidance, and contributions of several key individuals.

First, I would like to express my sincere gratitude for the time and effort of Dr Erik J. Sánchez for his insight and feedback regarding the state of this project as it was being developed. FDTD is an expansive and complex topic that can be quite daunting upon introduction, Dr. Sánchez's expertise introduced me to many aspects of the work I was able to demonstrate in this paper.

Additionally, I would also like to pay special acknowledgment to the work of Dr. John B. Schneider from the School of Electrical Engineering and Computer Science, Washington State University. Dr. Schneider's outstanding textbook "Understanding the Finite-Difference Time-Domain Method" [12] has been an immensely helpful resource along my journey through this topic and has greatly inspired my approach to the material covered within this paper.

References

[1] Kevin Cahill. *Physical Mathematics*. 2nd. Cambridge: Cambridge University Press, 2019. ISBN: 978-1-108-47003-2. DOI: 10.1017/9781108555814. URL: https://www.cambridge.org/9781108470032.

[2] Richard P. Feynman, Robert B. Leighton, and Matthew Sands. *The Feynman Lectures on Physics, Vol. II*. New Millennium Edition. New York, NY: Basic Books, a subsidiary of Hachette Book Group, 2011. ISBN: 978-0-465-02416-2. URL: https://www.basicbooks.com/.

[3] David J. Griffiths. *Introduction to Electrodynamics*. 4th ed. Glenview, IL: Pearson, 2013. ISBN: 978-0-321-85656-2.

[4] John David Jackson. *Classical Electrodynamics*. 3rd. Hoboken, NJ: John Wiley & Sons, 1998. ISBN: 978-0-471-30932-1.

[5] Karl S. Kunz and Raymond J. Luebbers. *The Finite Difference Time Domain Method for Electromagnetics*. Boca Raton, Florida: CRC Press, 1993. ISBN: 0-8493-8657-8.

[6] Antonino Calà Lesina et al. "On the convergence and accuracy of the FDTD method for nanoplasmonics". In: *Opt. Express* 23.8 (Apr. 2015), pp. 10481–10497. DOI: 10.1364/OE.23.010481. URL: https://opg.optica.org/oe/abstract.cfm?URI=oe-23-8-10481.

[7] Antonio Calà Lesina. *Silver plasmonic spheres on top of silver by parallel 3-D FDTD on IBM BlueGene/Q. Lateral view.* Online; A YouTube video depicting the output of a FDTD simulation by the channel "@AntonioCalaLesina". Online; available https://www.youtube.com/watch?v=r-HWbAmcnKo. June 2015. URL: \url{https://www.youtube.com/watch?v=r-HWbAmcnKo}.

[8] LibreTexts Chemistry. *The Nyquist Sampling Theorem*. Online; available: https://chem.libretexts.org/Ancillary_Materials/Laboratory_Experiments/Wet_Lab_Experiments/Analytical_Chemistry_Labs/ASDL_Labware/Analog_and_Digital_Conversion_for_Chemical_Instrumentation/04_Number_Representation/05_The_Nyquist_Sampling_Theorem. June 2020. URL: \url{https://chem.libretexts.org/Ancillary_Materials/Laboratory_Experiments/Wet_Lab_Experiments/Analytical_Chemistry_Labs/ASDL_Labware/Analog_and_Digital_Conversion_for_Chemical_Instrumentation/04_Number_Representation/05_The_Nyquist_Sampling_Theorem}.

[9] James R. Nagel. *The One-Dimensional Finite-Difference Time-Domain (FDTD) Algorithm Applied to the Schrödinger Equation*. Available https://scipy-cookbook.readthedocs.io/_downloads/bca7b26f2efcc85a15a70187c3989aa1/Schrodinger_FDTD.pdf.

[10] Gernot Ohner. "Two Dimensional Finite Difference Time Domain Computation of Electromagnetic Fields in Python". Bachelor's thesis. Graz: Karl-Franzens-Universität Graz, Mar. 2018.

[11] Raymond C. Rumpf. *Electromagnetic Analysis Using Finite-Difference Time-Domain*. Online; A series of YouTube video lectures on the FDTD published by the channel "@EMPossible". Available: https://www.youtube.com/playlist?list=PLLYQF5WvJdJWoU9uEeWJ6-MRzDSziNnGt. Nov. 2020. URL: \url{https://www.youtube.com/playlist?list=PLLYQF5WvJdJWoU9uEeWJ6-MRzDSziNnGt}.

[12] John B. Schneider. *Understanding the Finite-Difference Time-Domain Method.* Available https://eecs.wsu.edu/~schneidj/ufdtd/. 355 NE Spokane St, Pullman, WA 99163: School of Electrical Engineering and Computer Science, Washington State University, Aug. 2023.

[13] Allen Taflove and Susan Hagness. *Computational Electrodynamics: The Finite-Difference Time-Domain Method.* 2nd. Vol. 67-106. 2000. ISBN: 1-58053-076-1.

[14] UC Berkeley EECS Department. *The Nyquist-Shannon Sampling Theorem.* Available: https://ptolemy.eecs.berkeley.edu/eecs20/week13/nyquistShannon.html. 2020. URL: \url{https://ptolemy.eecs.berkeley.edu/eecs20/week13/nyquistShannon.html}.

[15] Kane Yee. "Numerical solution of initial boundary value problems involving maxwell's equations in isotropic media". In: *IEEE Transactions on Antennas and Propagation* 14.3 (1966), pp. 302–307. DOI: 10.1109/TAP.1966.1138693.

Utilization of ZIF-8 for cation adsorption and heavy metal ion filtration in mining effluents

Giancarlo Quispe[1], Antonio Lazo[1], Marcelo Díaz[2], Jean Rodriguez[2, 3] and Luis Acuña[4]

[1]Faculty of Geological, Mining and Metallurgical Engineering, National University of Engineering, Av. Tupac Amaru 210, Rímac, Lima, Peru
[2]Faculty of Physics, National University of San Marcos, Av. Carlos Germán Amezaga 375, Cercado de Lima, Lima, Peru
[3]Faculty of Sciences, National University of Engineering, Av. Tupac Amaru 210, Rímac, Lima, Peru
[4]Faculty of Industrial Engineering and Systems, National University of Engineering, Av. Tupac Amaru 210, Rímac, Lima, Peru

Abstract. The study of zeolites has aroused a growing interest due to their physical and chemical properties that are used in the refining of hydrocarbons in the oil industry, as well as in the removal of heavy metals, in the mining industry. The present work focused on the evaluation of the adsorption capacity of ZIF-8 in cations of Zn, Mn, Mg and other metals present in mining tailings water. Finally, the adsorption values obtained were between 33.88% (for boron) up to 98.88% (for arsenic), these results being compared by two other methods.
Keywords: Zeolites, zeolitic imidazolate framework, adsorption, filtered.
E-mail: alazo@uni.edu.pe

1. Introduction

Zeolitic imidazolate frameworks are comparable to zeolites, Niknam M. [1], which are hydrated aluminosilicates that occur in crystalline structures, Xu R. [2], among other similarities. Such as having relatively large cavities (in nanometers), Zhang H.[3], as well as ion exchange, Veerapandian S.[4], and reversible dehydration. On the other hand, zeolites are naturally stable, and are capable of withstanding environmental conditions that affect other materials, Park K.[5]. In the present work, the compound ZIF-8 (related to zeolites) was synthesized, to then proceed to compare its ability to filter elements present in water from mining tailings, with other similar methods, Rimarachin P.[6], Porras G.[7].

2. Research Process

The steps for the synthesis of Zeolitic imidazolate framework-8 (ZIF-8) , Pan Y.[8], are the following:

- Make a solution of 1.17 g of zinc nitrate hexahydrate dissolved in 8 g of deionized water.
- Make a solution of 22.70 g of 2-methylimidazolium dissolved in 80 g of deionized water.
- Mix both solutions and shake for 5 minutes, at room temperature.
- Subject the mixture to centrifugation for 10 minutes and wash the solid with deionized water. Repeat this step about 4 times.
- Finally, it is dried at 65 °C, Huang A.[9], for 12 hours.

2.1. Characterization Techniques of ZIF-8

The characterization techniques used in this study were X-Ray Diffraction (XRD) and Inductively Coupled Plasma Emission Spectroscopy method (ICP), Díaz M.[10].

Journal of Physics: Conference Series **3052** (2025) 012003 doi:10.1088/1742-6596/3052/1/012003

2.2. General Problem
Neutralizing mining and metallurgical tailings is essential for the coexistence of mining and agricultural, livestock, etc. activities. For this reason, methods are sought to neutralize the harmful effects of tailings. These include the use of zeolites, or, as in this work, the use of zeolitic imidazolate frameworks due to the greater variety that these present.

2.3. General Objective
To evaluate the adsorption capacity of ZIF-8 in a cation solution extracted from mining tailings from the concentration plant of the Universidad Nacional de Ingeniería (UNI).

2.4. Specific Objectives

- To evaluate the adsorption of cations present in a solution extracted from mining tailings from the concentration plant of UNI by the ZIF-8.
- Make a comparative analysis of the adsorption of cations present in this solution, with respect to the activated carbon and humus with compost methods.

2.5. General Hypothesis
ZIF-8 is used in mining tailings to improve the absorption of a cation solution extracted from mining tailings, of the concentrator plant of UNI.

2.6. Specific Hypotheses

- ZIF-8 improves the adsorption of cations present in a solution extracted from mining tailings from the concentration plant of the UNI.
- And it improves too when is compared with the use of activated carbon and the use of humus with compost.

2.7. Study Variable
Percentage of adsorption of cations in the solution extracted from the mining tailings of the concentrator plant of the UNI, by the ZIF-8 compound.

2.8. Intervening Variable
Mass of the elements present in the solution extracted from the mining tailings of the concentrator plant of the UNI.

3. Theoretical Framework
One of the most useful features of zeolites is that the metal cations in their structure are replaceable by others, Calleja A.[11]. This is useful, for example, in the treatment of hard water, where calcium cations are replaced by sodium or potassium cations, or decontamination of mining tributaries, Osorio A.[12].

On the other hand, although in Peru there are no large deposits of zeolites, Costafreda J.[13], it does present this type of compounds.

Regarding its synthetic generation, there is a huge variety of synthetic zeolites, Linares C.[14], Zhou R.[15], Lazo A.[16], with more appearing year after year.

3.1. Physical Properties of Zeolite
The most important properties of zeolites, which can be very useful to us, are: porosity, ion exchange, Chávez S.[17], and adsorption, Flores J.[18], Dabrowski A.[19].

3.2 Zeolitic Imidazolate Framework (ZIF)
The Zeolitic Imidazolate Framework has a structure similar to zeolites, Kida K.[20], Banerjee R.[21].

Journal of Physics: Conference Series **3052** (2025) 012003 doi:10.1088/1742-6596/3052/1/012003

ZIFs have an important advantage over zeolites, which have greater flexibility regarding structural surface modification. Compared with other metal-organic frameworks, ZIFs show higher chemical and thermal stability. For this reason, these materials have attracted much attention.

4. Methodology

4.1. ZIF-8 Synthesis Procedure

After having synthesized ZIF-8, following the steps indicated previously, we will obtain as a product a white solid in powder form. As expected from ZIF-8.

Once synthesized, an XRD analysis is applied to it, in order to determine the presence of ZIF-8.

4.2. Tailings Material Preparation

A sample of the tailings from the concentrator plant of the UNI was collected. For this purpose, a square mesh of approximately 1 meter on each side was drawn, with five points, to later extract samples from these, making conical excavations between 20 cm and 25 cm deep, and an approximate diameter of half a meter (Figure 1).

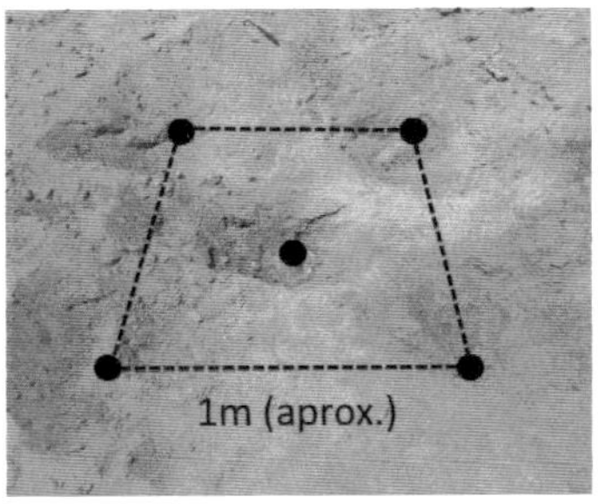

Figure 1. Five-point excavation for the tailings sample.

Once the material was extracted, it was placed in a plastic container and mixed with deionized water. It was then shaken and the plastic container was left in a cool and dark area for two months. To produce two well-defined phases: solid and liquid phase. The latter is the one we will use.

4.3. Filtering of Cation-Rich Solution from Tailings

After two months, a few milliliters of the liquid phase are extracted. Then 2 ml of tailings solution (rich in ions) was filtered: first through filter paper (kyntel brand 100 pcs) and then by ZIF-8 (Figure 2), supported by filter paper, using approximately 0.439 g of the compound.

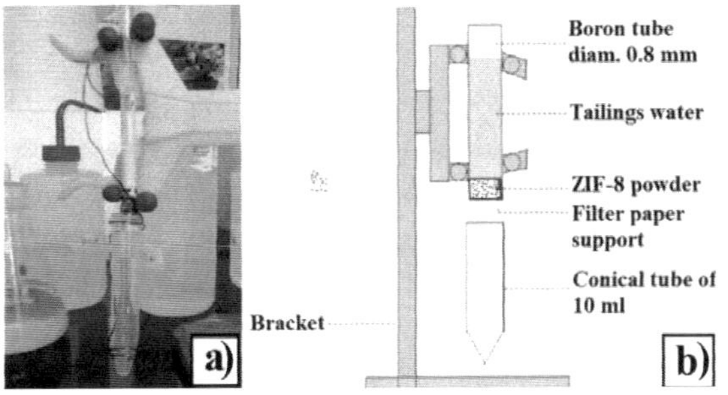

Figure 2. a) Photograph of the filtering implements. b) Schematic representation of the filtering process.

Both resulting liquids were diluted in 250 ml of deionized water, and then subjected to ICP analysis, to determine their chemical composition and the percentage of elements present in the samples after filtering with ZIF-8. These samples were labeled as follows: unfiltered tailings (solution L1), filtered with filter paper (solution L2), and filtered with ZIF-8 held on filter paper (solution L3).

5. Results

An XRD was performed on the sample, in order to identify the ZIF-8 compound (Figure 3).

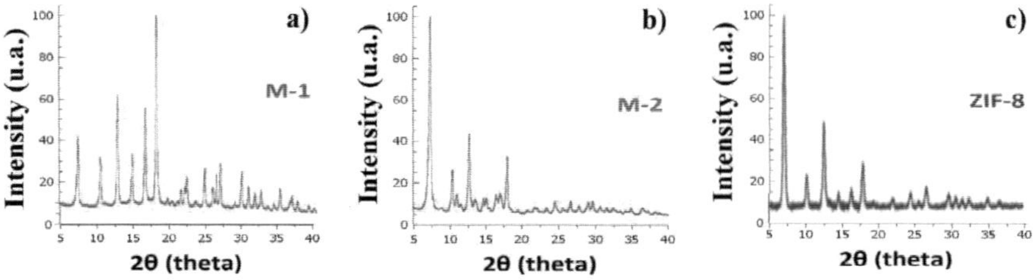

Figure 3. Comparison between the diffractograms a) of the first sample (M-1), b) of the second sample (M-2) and c) of the reference (ZIF-8), Pan Y.[8] .

The diffractogram of sample M-1 shows peaks that do not coincide with the X diffractogram of the reference, Pan Y.[8], as shown in Figure 3a. This concludes that ZIF-8 was not completely formed. Then, the sample was subjected to washing with ethanol. After this the substance obtained was identified as sample M-2, whose diffractogram is similar to the reference diffractogram (Figure 3b and 3c).

After the synthesis and characterization of ZIF-8, this material was used in the filtering of tailings. Table 1 shows the results of the ICP analysis on solutions L1 and L2.

Table 1. Content (in mg/L) of elements in a 2 ml solution sample. L1 represents the initial content, L2 the content after the use of filter paper and L1-L2 the content of elements retained by the filter paper.

Metal Assay	Sampling Limit (mg/L)	L1 (mg/L)	L2(mg/L)	L1-L2 (mg/L)
Manganese (Mn)	0.000010	8.202340	7.982440	0.219900
Silicate	0.010000	0.980000	0.850000	0.130000
Silica	0.008000	0.773000	0.672000	0.101000
Silicon (Si)	0.004000	0.361000	0.314000	0.047000
Zinc (Zn)	0.000050	42.635130	42.596680	0.038450
Magnesium (Mg)	0.004000	6.126000	6.090000	0.036000
Iron (Fe)	0.000050	0.046900	0.025920	0.020980
Lead (Pb)	0.000100	0.008900	0.003000	0.005900
Silver (Ag)	0.000020	0.004640	0.002270	0.002370
Boron (B)	0.000200	0.018000	0.015900	0.002100
Arsenic (As)	0.000010	0.002690	0.001140	0.001550
Bismuth (Bi)	0.000004	0.001517	0.000332	0.001185
Thallium (Tl)	0.000020	0.000840	0.000190	0.000650
Tungsten (W)	0.000020	0.000790	0.000210	0.000580
Molybdenum (Mo)	0.000050	0.000690	0.000140	0.000550
Cobalt (Co)	0.000006	0.029726	0.029276	0.000450
Thorium (Th)	0.000005	0.000521	0.000229	0.000292
Titanium (Ti)	0.000050	0.000420	0.000150	0.000270

Antimony (Sb)	0.000100	0.002400	0.002200	0.000200
Tantalum (Ta)	0.000010	0.000260	0.000100	0.000160
Cesium (Cs)	0.000020	0.000130	0.000040	0.000090
Lutetium (Lu)	0.000001	0.000024	0.000005	0.000019

The percentage of elements adsorbed by the filter paper is shown below (Table 2).

Table 2. Percentage of the elements retained by the filter paper of Table 1.

Metal Assay	%
Molybdenum (Mo)	79.71
Bismuth (Bi)	78.11
Thallium (Tl)	77.38
Tungsten (W)	73.41
Cesium (Cs)	69.23
Lead (Pb)	66.29
Titanium (Ti)	64.28
Tantalum (Ta)	61.53
Arsénic (As)	57.62
Thorium (Th)	56.04
Silver (Ag)	51.07
Iron (Fe)	44.73
Silicate	14.28
Silica	13.06
Silicon (Si)	13.01
Boron (B)	11.66
Antimony (Sb)	8.33
Lutetium (Lu)	4.16
Manganese (Mn)	2.68
Cobalt (Co)	1.51
Magnesium (Mg)	0.58
Zinc (Zn)	0.09

Table 3 shows the results of the ICP analysis on solutions L1 and L3, which correspond to the tailings before and after passing through the ZIF-8 (0.439 g powder). Where the amount of elements adsorbed by ZIF-8 (column L1-L3) is generally a large part, or almost all, of the elements present (column L1).

Table 3. Content (in mg/L) of elements in a 2 ml solution sample. L1 represents the initial content, L3 the content after the use of filter paper and ZIF-8, and L1-L3 the content retained by the filter paper and ZIF-8.

Metal Assay	Sampling Limit (mg/L)	L1 (mg/L)	L3(mg/L)	L1-L3 (mg/L)
Antimony (Sb)	0.000100	0.002400	0.00000(*)	0.002400
Titanium (Ti)	0.000050	0.000420	0.00000(*)	0.000420
Zinc (Zn)	0.000050	42.635130	5.072850	37.562280
Manganese (Mn)	0.000010	8.202340	3.026850	5.175490
Magnesium (Mg)	0.004000	6.126000	2.412000	3.714000
Silicate	0.010000	0.980000	0.280000	0.700000

Silica	0.008000	0.773000	0.225000	0.548000
Silicon (Si)	0.004000	0.361000	0.105000	0.256000
Iron (Fe)	0.000050	0.046900	0.002290	0.044610
Cobalt (Co)	0.000006	0.029726	0.008065	0.021661
Lead (Pb)	0.000100	0.008900	0.000400	0.008500
Boron (B)	0.000200	0.018000	0.012900	0.005100
Silver (Ag)	0.000020	0.004640	0.001500	0.003140
Arsenicc (As)	0.000010	0.002690	0.000030	0.002660
Bismuth (Bi)	0.000004	0.001517	0.000261	0.001256
Thalium (Tl)	0.000020	0.000840	0.000090	0.000750
Tungsten (W)	0.000020	0.000790	0.000110	0.000680
Molybdenum (Mo)	0.000050	0.000690	0.000050	0.000640
Thorium (Th)	0.000005	0.000521	0.000079	0.000442
Tantalum (Ta)	0.000010	0.000260	0.000080	0.000180
Cesium (Cs)	0.000020	0.000130	0.000020	0.000110
Lutetium (Lu)	0.000001	0.000024	0.000001	0.000023

(*)The amount is too small to be quantified.

The percentage of elements absorbed by the ZIF-8 compound is shown below (Table 4).

Table 4. Percentage of the elements adsorbed by the mass of ZIF-8 (0.439 g), supported by filter paper.

Metal Assay	%
Antimony (Sb)	100.0(*)
Titanium (Ti)	100.0(*)
Arsenic (As)	98.88
Lutetium (Lu)	95.83
Lead (Pb)	95.50
Iron (Fe)	95.11
Molybdenum (Mo)	92.75
Thalium (Tl)	89.28
Zinc (Zn)	88.10
Tungsten (W)	86.07
Thorium (Th)	84.83
Cesium (Cs)	84.61
Bismuth (Bi)	82.79
Cobalt (Co)	72.86
Silicate	71.42
Silicon (Si)	70.91
Silica	70.89
Tantalum (Ta)	69.23
Silver (Ag)	67.67
Manganese (Mn)	63.09
Magnesium (Mg)	60.62
Boron (B)	33.88

(*)The amount not adsorbed by ZIF-8 is too small to quantify, therefore complete adsorption is assumed.

Table 4 shows that most of the elements are adsorbed up to 70% of their initial value. Table 5 show those elements for which the filter paper adsorbed scarcely, but were mostly adsorbed by the ZIF-8.

Table 5. Percentage of elements mostly adsorbed by ZIF-8. Compared to filter paper.

Metal Assay	Filter Paper (%)	ZIF-8 and Filter Paper (%)
Lutetium (Lu)	4.16	95.83
Zinc (Zn)	0.09	88.10
Cobalt (Co)	1.51	72.86
Manganese (Mn)	2.68	63.09
Magnesium (Mg)	0.58	60.62

Table 5 shows that ZIF-8 has a good adsorption capacity for zinc (88.1%), among other elements. The highest percentages are shown in Table 6.

Table 6. Higher percentages of adsorption (more than 85%) of the mass of ZIF-8 (0.439 g) held by the filter paper.

Metal Assay	%
Arsenic (As)	98.88
Iron (Fe)	95.11
Lutetium (Lu)	95.83
Molybdenum (Mo)	92.75
Lead (Pb)	95.50
Thallium (Tl)	89.28
Tungsten (W)	86.07
Zinc (Zn)	88.10

On the other hand, Figure 4 shows the percentage of lead removed from a sample of mining tailings using the activated carbon method, Porras G.[7]. Which is capable of extracting a good percentage of lead, but the time spent on it (hours) could limit its application.

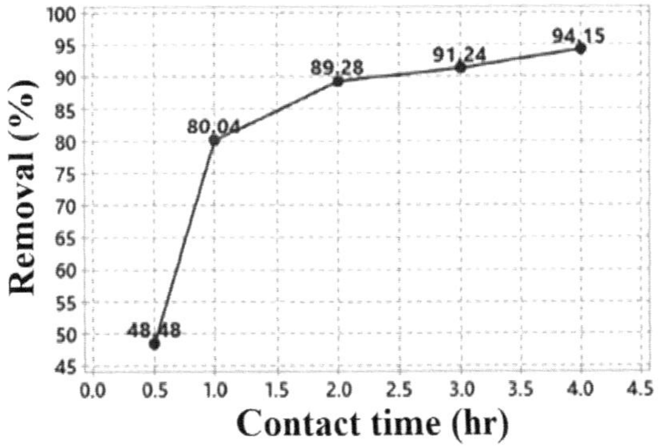

Figure 4. Percentage of lead adsorption over time using the activated carbon method, Porras G.[7].

For the ZIF-8, the filtering of elements in the tailings took 30 minutes, contrasting with the activated carbon method. In the case of the use of humus and compost, adsorption by ZIF-8 also has advantages:

Journal of Physics: Conference Series **3052** (2025) 012003 doi:10.1088/1742-6596/3052/1/012003

- The activated carbon method achieves a filtering percentage of the lead element lower than that of ZIF-8 (95.5%, compared to the adsorption of activated carbon, which reaches 94.15%).
- The commercial humus and compost method, Rimarachin P.[6], reaches a filtering percentage of the zinc element of 85.67%, in 15 days, unlike when the ZIF-8 compound is used, which reaches a percentage of 88.10% in 30 minutes.

Although it should be considered that the filtering carried out in this work was done in a laboratory.

6. Analysis of Results

From the ZIF-8 synthesis process, its structural characterization by X-ray diffraction and ICP analysis, the following is concluded:

- Washing with ethanol and centrifugation increases the proportion of ZIF-8 (evidenced by the white coloration acquired by the sample).
- It is with magnesium, manganese, cobalt, zinc and lutetium where the best results can be seen in terms of ZIF-8 adsorption.
- Taking into account that the mass of the ZIF-8 compound used in the tests was 0.439 grams, the percentages of adsorption and the ratio (of masses) can be obtained. For example (Table 7):

Table 7. Percentages, masses and adsorption ratios of ZIF-8 (0.439 grams), for elements of higher concentration in the solution.

Metal Assay	Adsorption (%)	Mass (mg)	Ratio (mg element / gr. ZIF-8)
Zinc (Zn)	88.10	9.390	21.389
Manganese (Mn)	63.09	1.293	2.947
Magnesium (Mg)	60.62	0.928	2.115
Silicate	71.42	0.175	0.398
Silica	70,89	0.137	0.312

- Zinc is the most abundant element in the tailings solution. And, as shown in Table 7, the ZIF-8 compound has considerable ability to filter elements this element. Making special mention of the arsenic element (Table 6), of which ZIF-8 was able to adsorb a percentage of 98.88.

7. Conclusions

Compound ZIF-8 was obtained with minimal spurious phases requiring ethanol washing after synthesis. And it shows a good capacity to remove elements such as: titanium, iron, zinc, arsenic, molybdenum, lutetium, tungsten, thallium and lead.

In a mining tailings rich in Zinc, the ZIF-8 absorbed a percentage of 88.1% of Zinc in 30 minutes.

ZIF-8 compound is a good candidate for the purification of water contaminated with arsenic.

For certain elements present in the tailings, the adsorption through the ZIF-8 presents better results compared to other methods. As in the case of activated carbon, or in the case of humus and compost.

Acknowledgment

We would like to thank the Facultad de Ingeniería Geológica, Minera y Metalúrgica of UNI for the facilities, equipment and reagents provided for the development of this applied research topic, useful for the care of the environment and the national market.

References

[1] Niknam M, Ghahramaninezhad M and Eydifarash M 2017 Zeolitic imidazolate framework-8 for efficient adsorption and removal of Cr(VI) ions from aqueous solution *Environmental Science and Pollution Research* **24** pp 9624-9634

[2] Xu R, Pang W, Yu J, Huo Q and Chen J 2007 *Chemistry of Zeolites and Related Porous Materials: Synthesis and Structure* ed John Wiley & Sons pp 1-3

[3] Zhang H, Zhong J, Zhou G, Wu J, Yang Z and Shi X 2016 Microwave-assisted solvent-free synthesis of zeolitic imidazolate framework-67 *Jornal of Nanomaterials* **2016** 9648386

[4] Veerapandian S, De Geyter N, Giraudon J, Lamonier J, and Morent R 2019 The use of zeolites for VOCs abatement by combining non-thermal plasma, adsorption, and/or catalysis *Catalysts* **9** p 98

[5] Park K, Ni Z, Cote A, Choi J, Huang R, Uribe-Romo F, Chae H, O'Keeffe M and Yaghi O 2006 Exceptional chemical and thermal stability of zeolitic imidazolate frameworks *PNAS* **103**

[6] Rimarachin P and Huaranga F 2015 Tratamiento de aguas de efluentes minero–metalúrgicos utilizando, métodos pasivos y activos en sistemas experimentales *Sciéndo* **18**

[7] Porras G, Martinez C, Pampa N 2019 *Adsorción del Plomo Mediante Carbón Activo de Coronta de maíz de aguas contaminadas con relaves mineros*. (Lima: Universidad Peruana Unión) p 73

[8] Pan Y, Liu Y, Zeng G, Zhao L and Lai Z 2010 Rapid synthesis of zeolitic imidazolate framework-8 (ZIF-8) nanocrystals in an aqueous system *Chemical Communication* **47** pp 2071-2073

[9] Huang A and Caro J 2011 *Advances in Induction and Microwave Heating of Mineral and Organic Materials* ed Stanisław Grundas pp 641-672

[10] Diaz M and Lazo A 2022 Synthesis by hand milling and structural, porous and superficial characterization of the zeolitic imidazolate framework-67 *Journal of Physics: Conference Series* **2678** 012002

[11] Calleja A 2009 La Importancia de las Zeolitas *Universidad de Salamanca* CT 1 (2009) 211-227

[12] Osorio A, Bazán D, Carhuancho M, Salas N, Zárate G, Lengua R, Aguirre R, Becerra E 2004 Aplicación de la montmorillonita en la descontaminación de afluentes mineros *Rev. Soc. Quim. Perú* **70** pp 18-26

[13] Costafreda J, Martin D, Costafreda V. J, Prado R, Iván J, Álvarez Y, Bello L, Vattuone M, Gargiulo M, Crosta S, Rodrigo P, Rosell M, Magalhaes C, Benvindo A, Morante F, Novo R, Carrasco R and Gajardo A 2016 *Las Zeolitas Naturales de Iberoamerica* ed Fundación Gomez-Pardo (Madrid: Universidad Politécnica de Madrid)

[14] Linares C, Goldwasser M, Machado F and Rodriguez G 1999 Zeolitas sintéticas versus zeolitas naturales como catalizadores básicos *Sociedad Venezolana de Química* **22** pp 9-13

[15] Zhou R, Zhong S, Lin X, Xu N 2009 Synthesis of zeolite T by microwave and conventional heating, microporous and mesoporous materials *Elsevier* **124** pp 117-122

[16] Lazo A, Yparraguirre J and Sedano R 2019 *Síntesis y caracterización de Zeolita (LTA) y de marcos zeoliticos de Imidazolato (ZIF-67) por el método de microondas* (Lima Universidad Nacional de Ingeniería)

[17] Chávez S, Canoira L and Costafreda J 2017 *Ensayos de Intercambio Cationico de Zeolitas Naturales con Elementos de Tierras Raras* (Madrid: Universidad Politécnica de Madrid)

[18] Flores J, Maubert A and Martín N 2006 Evaluation of the exchange properties of a natural mexican zeolite for N2-O2 separation in the atmospheric Air *Revista Mexicana de Ingenieria Quimica* **5** pp 119-129

[19] Dabrowski A 2001 Adsorption-from theory to practice *Elsevier* **93** pp 135-224

[20] Kida K, Okita M, Fujita K, Tanaka S and Miyake Y 2013 Formation of high crystalline ZIF-8 in an aqueous solution *CrystEngComm* **15** pp 1794-1801

[21] Banerjee R, Phan A, Wang B, Knobler C, Furukawa H, O'Keeffe M and Yaghi O 2008 High-throughput synthesis of zeolitic imidazolate frameworks and application to CO_2 capture *Science* **319** pp 939-943

A Model for the catalytic reduction of CO on a surface with variable reaction rate, CO desorption and impurities

G. J. Pinto[1] **and G. M. Buendía**[1]

[1] Physics Department, Universidad Simón Bolívar, Caracas 1080, Venezuela.

E-mail: buendia@usb.ve

Abstract.
Kinetic Monte Carlo simulations are applied to study a modified Ziff-Gulari-Barshad (ZGB) model for the catalytic oxidation of carbon monoxide on a surface. The original model has been made more realistic by incorporating temperature effects such as desorption of the CO, a variable reaction rate probability, and the existence of contaminants in the feeding gas. Several of these effects have already been studied separately, in this work we consider how they alter the system when considered simultaneously. We found that the combination of these processes strongly modifies the phase diagram of the original model, in many cases eliminating the abrupt phase transition towards a nonreactive state. And although the lessening of the reaction probability and the presence of impurities clearly reduce the productivity of the system, these effects can be counteracted by the CO desorption and obviously by the impurity desorption.

Keywords: ZGB Model, heterogeneous catalysis, non-reactive impurities.

1. Introduction

Catalytic reactions are a very relevant topic of study, not only for their fundamental role in living processes, environmental protection and industrial applications [1–3], but also because they constitute an example of out of equilibrium processes. Most systems in nature are out of equilibrium and present a large number of interesting phenomena such as chaoticity, new patterns, self-organization, irreversible phase transitions, etc. Due to the complexity and irreversibility of these processes there is not a well established general method to study them. Numerical simulations of models that describe heterogeneous catalytic reactions on surfaces have become an important tool to explore complex phenomena. The Ziff-Gulary-Barshad (ZGB) [4] model describes several aspects of the catalytic oxidation of CO on a surface, a fundamental process associated with the control of harmful emissions from combustion engines [5]. In this model the catalytic surface is represented by a lattice in two dimensions in contact with an atmosphere that contains CO and molecular O_2, that are adsorbed on the lattice sites with probability y and $1 - y$ respectively, proportional to their partial pressures in the atmosphere. The CO adsorption requires one site while the O_2 requires two sites that are nearest-neighbors (nn) on the lattice. Adsorbed CO and O located in nn sites always react (probability $p = 1$) and liberate CO_2 to the atmosphere, leaving two empty sites on the lattice. This deceptively simple model that has only one parameter, y, in addition to describing some qualitatively aspects of the real catalytic process, has a rich behavior that includes a continuous phase transition at y_1 and a discontinuous one at $y_2 > y_1$, and presents adsorbing states, metastable states, etc [4, 6–9]. One of the advantages of this model is that it can be easily modified to include physical aspects such as temperature effects, that can be introduced as desorption, diffusion, or changes in the reaction rate [10–15]. Since contaminants are an unavoidable part

Journal of Physics: Conference Series **3052** (2025) 012004 doi:10.1088/1742-6596/3052/1/012004

of industrial processes, they can be included as part of the components of the atmosphere in contact with the surface [16–18]. Several authors have introduced these aspects in different ways, frequently studying the effect of each one separately. In this work we study the effect of introducing simultaneously several of these modifications. We change the original ZGB adding CO desorption, that can be considered as an effect associated with the temperature [19–22] (O desorption is not considered since experiments indicate that it is not relevant [19]), we also consider that the adsorbed CO and O react with a probability $p \leq 1$, also a temperature-related effect [19]. Our model also includes impurities in the atmosphere that can be adsorbed or desorbed on the surface. The original ZGB model has a continuous transition between a poisoned O, non-reactive phase, to a reactive phase at y_1, this transition has not been observed in nature, and one way to eliminate it is by changing the adsorption mechanism of the O_2 on the surface, such that instead of being adsorbed in two nn sites the oxygens are adsorbed in two next-nearest neighbors (nnn) sites [23], this change was inspired on experimental results that indicate that the O's move further apart once they are adsorbed in what has been called a hot atom mechanism [24], so along this work we will consider that the O_2 adsorb in two sites that are nnn on the lattice.

This paper is organized as follows. In Sec.II the modified ZGB model is described. In Sec.III the numerical results are presented in two different stages, we start by including separately the CO desorption and the variable reaction rate, $p \leq 1$, next we study the effect of adding non-desorbing impurities to the atmosphere combined to the previous temperature effects. Finally we present our conclusions.

2. The model

The catalytic surface is simulated by a square lattice of size LxL. In our modified model, the lattice is in contact with an atmosphere that contains CO, O_2, and impurities X. A Kinetic Monte Carlo simulation [25] generates a sequence of trials: CO, O_2 or X adsorptions, or CO or X desorptions. A single site i is selected randomly, if it is filled, we try desorption: CO with probability k_{CO}, X with probability k_X, if it is empty, we try adsorption: CO with probability y, X with probability y_X, and O_2 with probability $1 - y - y_X$. The O_2 adsorption needs an extra empty site, nnn to the selected site i [23]. If there is more than one empty nnn site available, one of them is chosen at random. When an adsorption of CO or an adsorption of O_2 occurs, nn pairs are checked randomly. Pairs formed by nn CO and O react with probability $p \leq 1$, liberating a CO_2 molecule and leaving behind two nn empty sites on the lattice. The reactions that define the model are given by:

$$\begin{aligned}
\mathrm{CO}(g) + \mathrm{S} &\rightarrow \mathrm{CO}(a) \\
\mathrm{O}_2(g) + 2\mathrm{S} &\rightarrow \mathrm{O}(a) + \mathrm{O}(a) \\
\mathrm{CO}(a) + \mathrm{O}(a) &\rightarrow \mathrm{CO}_2(g) + 2\mathrm{S} \\
\mathrm{CO}(a) &\rightarrow \mathrm{CO}(g) + \mathrm{S} \\
\mathrm{X}(g) + \mathrm{S} &\rightarrow \mathrm{X}(a) \\
\mathrm{X}(a) &\rightarrow \mathrm{X}(g) + \mathrm{S}
\end{aligned} \tag{1}$$

S indicates an empty site on the lattice, g indicates gas phase and a indicates adsorbed. The first three steps define a Langmuir-Hinshelwood [26] mechanism that describes the original ZGB model (with the difference that in the ZGB model the third reaction, between the CO and the O adsorbed in nn sites, always occurs while in the present model it occurs with probability $p \leq 1$). The fourth, fifth and sixth steps correspond to the extra processes we are including: CO desorption, X adsorption and X desorption, respectively. We impose periodic boundary conditions and take $L = 120$. We define the unit of time as one Monte Carlo step per Site (MCSS), in which each site is visited on average once. Measurements are calculated with 10^5 MCSS after 5×10^4 MCSS are discarded to ensure that the system

Journal of Physics: Conference Series 3052 (2025) 012004 doi:10.1088/1742-6596/3052/1/012004

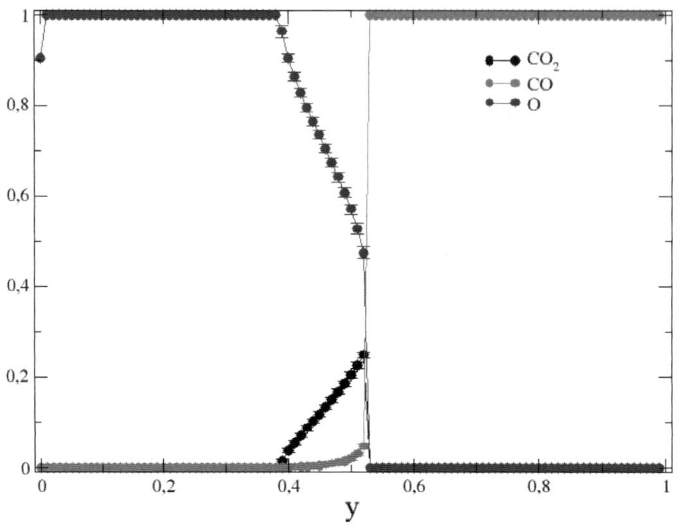

Figure 1: Coverages of the adsorbed species, CO and O, and rate of production of CO_2, vs y, for the original ZGB model. The system has a continuous transition between a non-productive and a productive region, at $y_1 \approx 0.39$ and a abrupt transition, at $y_2 \approx 0.52$ towards a non-productive phase where the lattice is poisoned with CO.

has reached the stability. We define coverage as the fraction of sites on the surface filled by an adsorbate. We measured CO, O, and X coverages, and the rate of production of CO_2 (also normalized by the number of sites). We always start from an empty lattice and wait until the system reach a steady state before taking any measurements.

The original ZGB model, i.e., the first three steps of Eq. (1) with $p = 1$, when simulated on a square lattice has two transitions, a continuous one at $y_1 \approx 0.39$ between a non-reactive state where the lattice is poisoned with O and a reactive state, and a sharp transition at $y_2 \approx 0.52$ between a reactive state and a non-reactive state where the lattice is poisoned with CO, Fig.1. As we already mentioned, the hot-atom mechanism introduced in the entrance adsorption of the O_2 eliminates the nonphysical continuous transition at y_1, such that now the system starts to produce CO_2 as soon as $y > 0$ until $y = y_2$, as seen Fig.2 where we plot the two rates of production.

3. Numerical results

3.1. ZGB model with CO desorption ,k_{CO}, and ZGB with variable reaction rate, $p \leq 1$

First we include only CO desorption, that is adding the step four to the first three of Eq.1, keeping the reaction rate $p = 1$. Now if a site i selected at random is filled with a CO we try desorption with probability k_{CO} [19, 27–29]. In Fig.3 we observe that the CO desorption not only increases the peak of production, but also widens the production window which now extends almost to the complete range of values of y. The coverages and production smooth out, and for k_{CO} large enough the sharp transition at y_2 disappears. As expected, the CO desorption is more relevant when $y \geq y_2$ where the lattice is mostly filled with CO, its desorption favors the eventual adsorption of O_2 that upon entrance keeps the reactions going on.

Next we change the reaction rate p while maintaining $k_{CO} = 0$. In the standard ZGB model nn CO and O always react, $p = 1$ and liberate CO_2, now this reaction occurs with probability $p \leq 1$. The experiments indicate that, within certain ranges, p decreases as the temperature decreases [19]. Our simulations show that there is a critical value of p, such that if $p \leq p_c \approx 0.520$ the abrupt transition at

Journal of Physics: Conference Series 3052 (2025) 012004 doi:10.1088/1742-6596/3052/1/012004

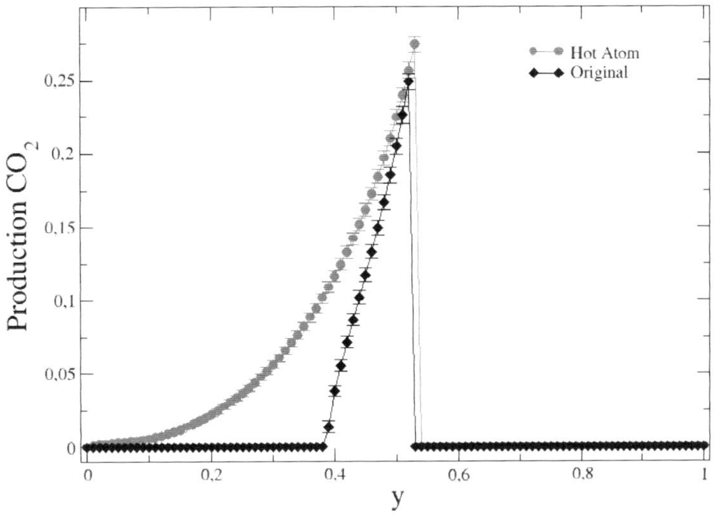

Figure 2: Comparison between the production rates from the original ZGB model and those obtained after modifying it by applying the "Hot Atom" mechanism where the O_2 dissociates upon adsorption. In the modified model, the unphysical continuous transition at y_1 disappears, the system produce CO_2 as soon as $y > 0$ until the discontinuous transition occurs.

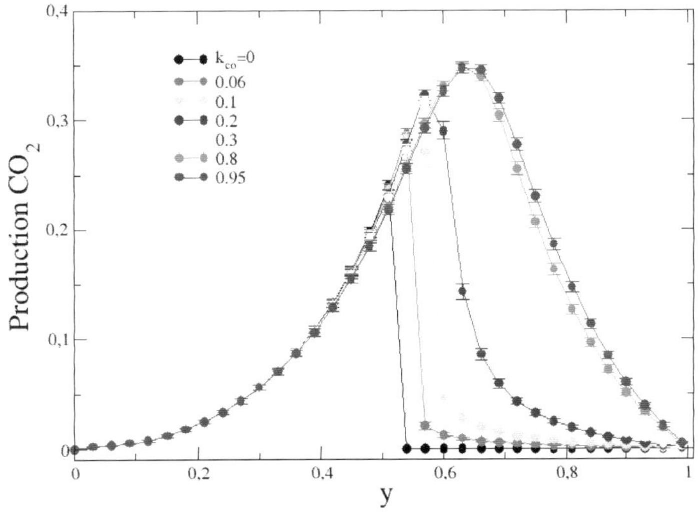

Figure 3: Rate of production of CO_2 vs y for different values of CO desorption k_{CO} and $p = 1$.

y_2 disappears, the lattice fills slowly with CO as y increases, while the O coverage dismiss slowly, see Fig.4a, 4b. Eventually for p small enough the lattice get filled with CO and O (in smaller amounts) such that there is no production. The production rate is presented in Fig.4c. As p is reduced the peak of the production and the window of production become smaller, until they eventually disappear. Notice that while the k_{CO} affects mostly the behavior of the system in the region $y \geq y_2$ the reduction of p affects it when $y \leq y_2$.

Journal of Physics: Conference Series **3052** (2025) 012004
doi:10.1088/1742-6596/3052/1/012004

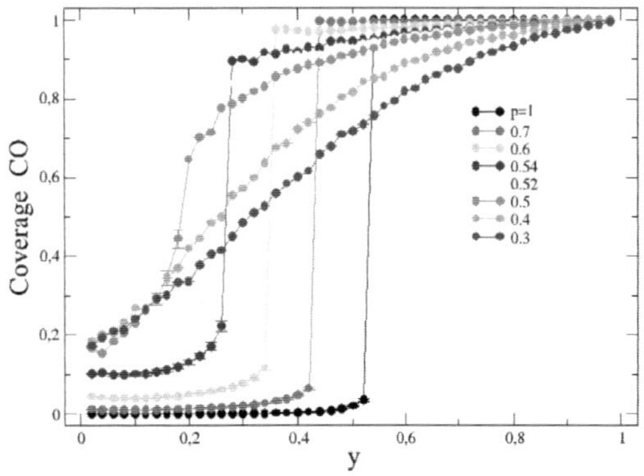

(a)

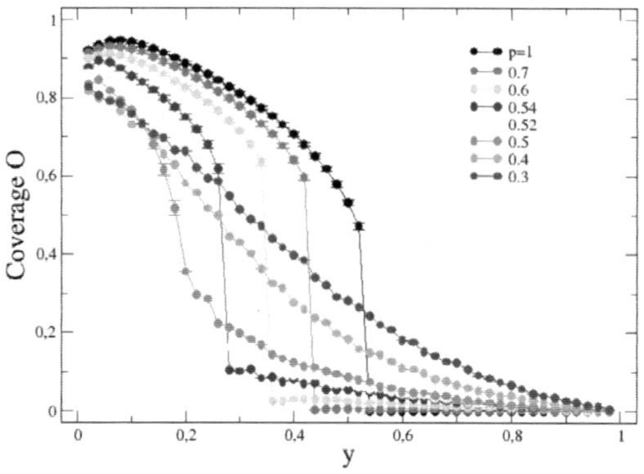

(b)

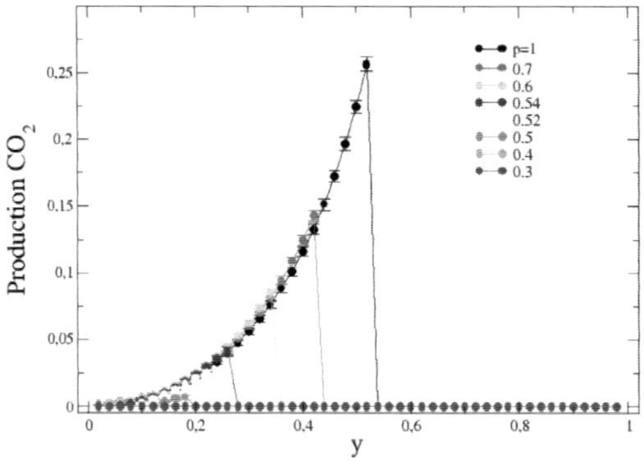

(c)

Figure 4: (a) CO coverage, (b) O coverage, and (c) CO_2 production rate, vs y for several values of p and $k_{CO} = 0$.

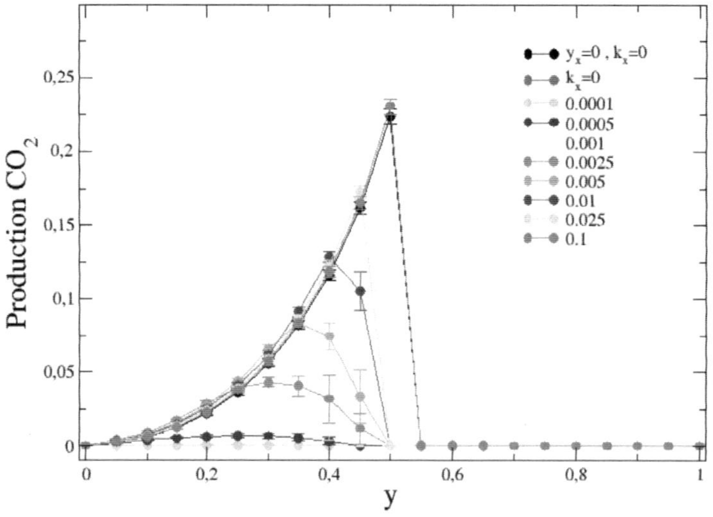

Figure 5: Rate of production of CO_2 vs y for several values of X desorption, k_X, and $k_{CO} = 0.005$, $p = 1$.

3.2. ZGB model with CO desorption, $k_{CO} \neq 0$, variable reaction rate, $p \leq 1$ and nonreactive impurities in the atmosphere, ZGB+ model

Here we analyze the effect of the presence of impurities in the gas phase to the previous models that included CO desorption and variable reaction rate between the CO and the O, we call this new model ZGB+, and it is defined by the six steps of Eq.1. Now the surface is in contact with an atmosphere that contains CO, O_2, and X, where X are non-reactive impurities that can be adsorbed and desorbed on the lattice. A site i is selected at random, if i is vacant we try to adsorb a CO, X or O_2 with probabilities y, y_X and $1 - y - y_X$, respectively, if it is occupied we try to desorb: CO with probability k_{CO}, X with k_X. Previous studies with this model with $p = 1$, i.e., the reaction between nn CO and O always occurs, indicate that, even when CO desorption is included, if the gas contains non-desorbing impurities, $k_X = 0$, it takes a long time for the system to become stable and in this case does not have a production rate. The lattice eventually get filled with X [29]. When the impurities can desorb the system can have a stable productive phase. In Fig.5 where y_X and k_{CO} are fixed, and $p = 1$ we see that the production of CO_2 is zero for $k_X = 0$, increases as k_X increases until it reaches a peak and then decreases in a way that becomes more and more abrupt, until for large enough k_X the system behaves as the original ZGB model with CO desorption, in this case the effect of the impurities is almost negligible.

To analyze the ZGB+ model that includes all the effects mentioned, we start by fixing the amount of impurities in the gas, y_X, their desorption rate k_X, the desorption rate of the CO, k_{CO} and change the reaction probability p. The coverages are shown in Fig.6. Notice that as p decreases the coverages changes more smoothly with y. In the limit $y \to 1$ the lattice is mostly filled with CO with a few X and even fewer O, and the coverages become independent of p. As p keep decreasing, the coverage of X reaches a peak that decreases in height and moves towards lower values of y. Notice that the productivity, Fig.7, follows the same pattern as the X coverage. Comparing with the plots for the productivity for the cases where only CO desorption were included, Fig.3, and when only p variable were considered Fig.4c, we notice that now, due to the impurities and the combination of temperature effects, the changes are observed in the whole range of values of y. The coverages and the productivity now behave in a very smooth way, no abrupt transition. The impurities are relevant particularly for intermediate values of y. Also we notice that now the system remains productive even for relative low values of p, this is due to presence the CO desorption. Also the maxima of the production is lower than in the previous cases, due to the presence of the impurities.

Next we set y_X, k_X and $p < 1$ and change k_{CO}. The coverages are shown in Fig.8, we observe that they change more smoothly with y as k_{CO} increases. The X coverage, Fig.8c, is almost independent on k_{CO} for low y, then as y increases it increases almost linearly with k_{CO} until it reaches a peak and then decreases, the decrease becomes very slow for y and k_{CO} large, where the lattice is filled mostly with impurities and some CO. This indicates that the empty sites left from the high CO desorption are mostly filled by impurities. Again it is interesting that the plot of the production rates Fig.9, is very similar in shape to the one for the X coverage before the peak of production is reached, after which the decay in productivity is much faster and eventually there is no production of CO_2. As expected, the effect of the CO desorption are almost negligible for very low values of y. Again it is remarkable that as k_{CO} increases the range of production of the system expands so that it extends to all values of y.

Finally we set y_X, k_{CO} and $p < 1$ and change k_X, in Fig.10 we see that the productivity increases almost linearly with k_X, reaches a maximum and then decreases more and more abruptly until the system behaves as the original model recovering the abrupt transition at y_2. The system remains productive almost in the whole range of $y \leq y_2$.

Comparing the different curves, we see how some effects are counteracted while others are reinforced. Obviously the existence of impurities and the reduction in the reaction rate lower the productivity while the desorptions increase it. Their combined effects in most of the cases eliminate the abrupt transition that has the original model at y_2, and makes the system productive in the whole range of values of y. The k_{CO} counteracts the negative effect of the reduced p, while the k_X counteracts the effect of the impurities. Of course the global behavior of the system depends on the combined value of the different

Journal of Physics: Conference Series **3052** (2025) 012004 doi:10.1088/1742-6596/3052/1/012004

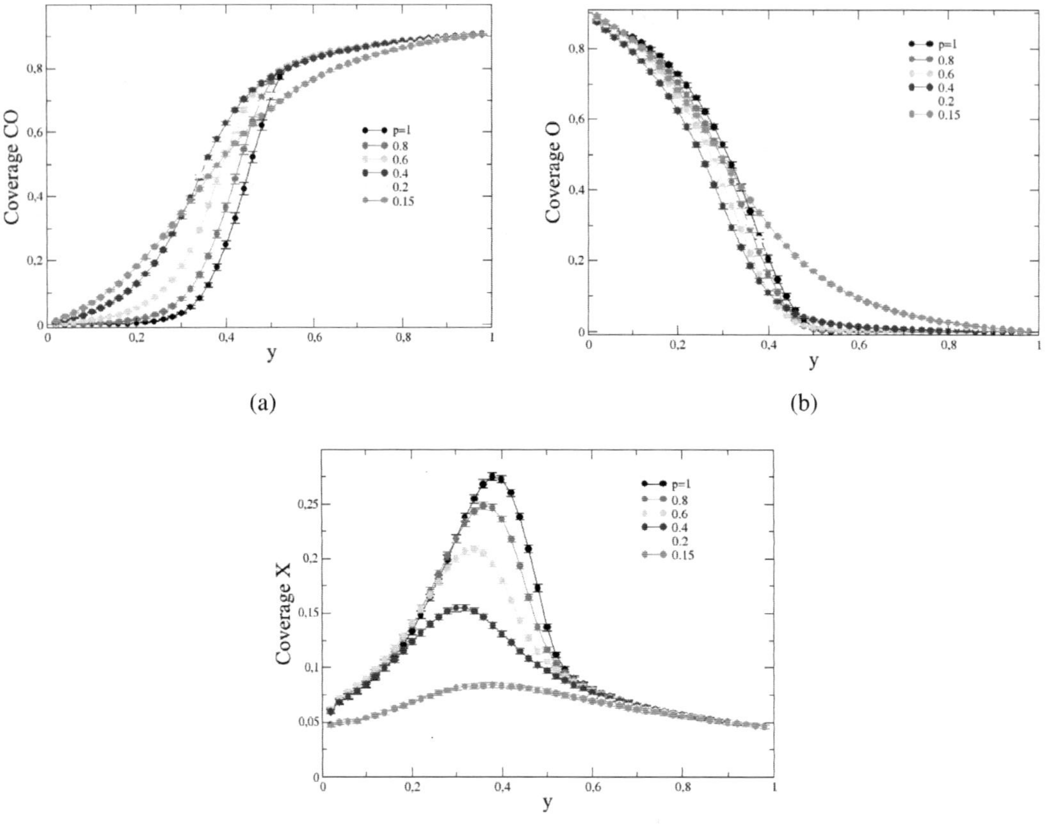

(a) (b) (c)

Figure 6: Coverages vs y for different values of p, $k_{CO} = 0.05$, and $y_X = k_X = 0.005$, ZGB+ model. (a) CO coverage, (b) O coverage, and (c) X coverage.

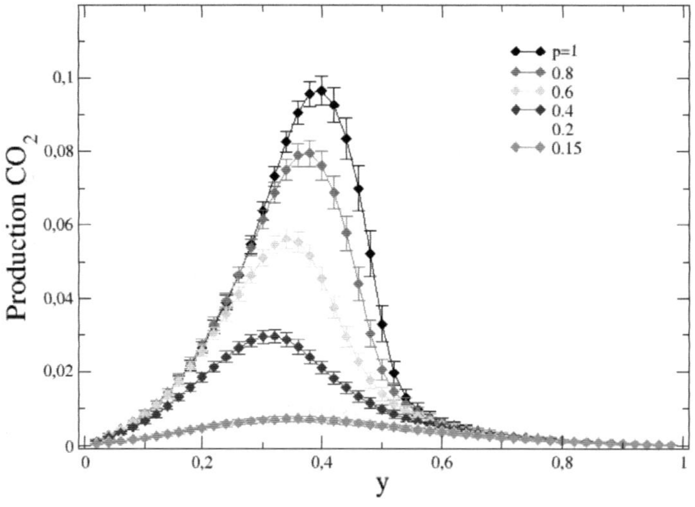

Figure 7: Rate of production of CO_2 vs y for different values of p, $k_{CO} = 0.05$, $y_X = k_X = 0.005$, ZGB+ model.

Journal of Physics: Conference Series **3052** (2025) 012004
doi:10.1088/1742-6596/3052/1/012004

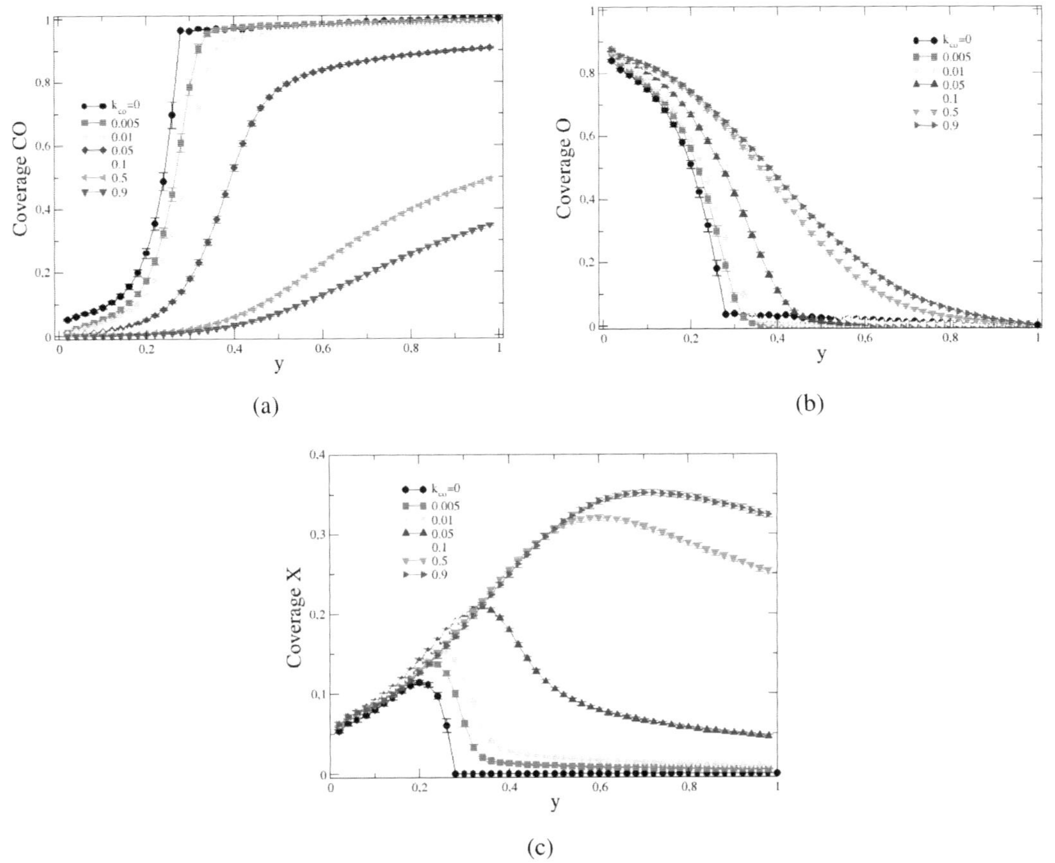

Figure 8: Coverages vs y for different values of k_{CO}, $p = 0.6$, and $y_X = k_X = 0.005$, ZGB+ model. (a) CO coverage, (b) O coverage, and (c) X coverage.

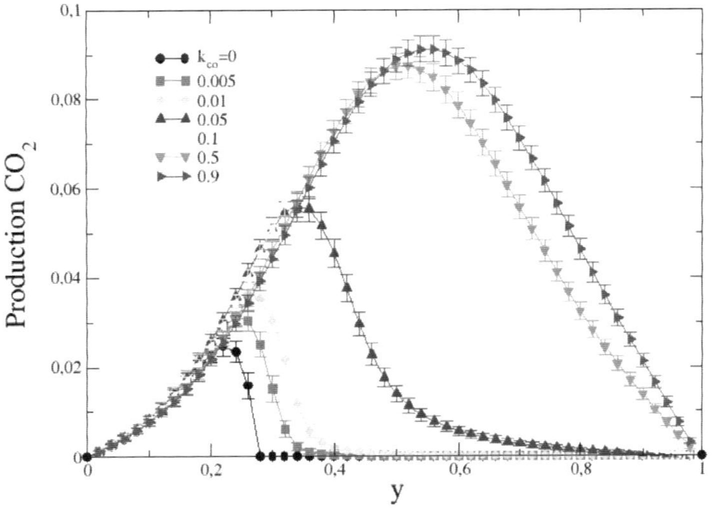

Figure 9: Rate of production of CO_2 vs y for several values of k_{CO}, $p = 0.6$, and $y_X = k_X = 0.005$, ZGB+ model.

Journal of Physics: Conference Series **3052** (2025) 012004 doi:10.1088/1742-6596/3052/1/012004

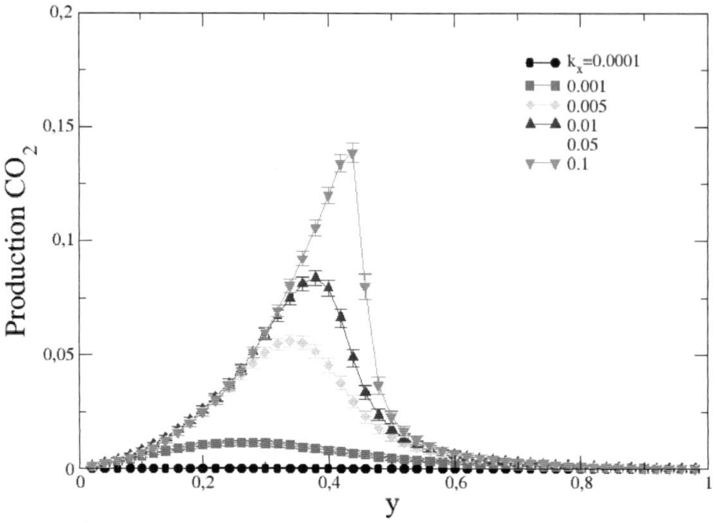

Figure 10: Rate of production of CO_2 vs y for different values of k_X, $p = 0.6$, and $y_X = k_{CO} = 0.005$, ZGB+ model.

parameters.

In Fig.11 we present some snapshots of the surfaces once the system has reached a stable state. In the figures, y, y_X and k_X remain fixed and k_{CO} and p change. The number of empty spaces (white dots) is an indicator of the productivity of the surface. Comparing Fig.11a and Fig.11b, both taken at the same value of $p = 0.6$, we can see the dramatic effect of changing the CO desorption, in both cases, even for the relative high value of $y \geq y_2$ and the reduced value of p, is evident that the surfaces are not in an adsorbing state, there are still empty spaces where adsorbates can enter and produce reactions. When k_{CO} is small Fig.11b, most of the lattice is filled with CO with very few oxygens and impurities, for a large k_{CO} Fig.11a the surface is mostly filled with empty spaces, oxygens and impurities. The surfaces change dramatically when p is strongly reduced. Comparing Fig.11c and Fig.11d, both taken for $p = 0.1$, now when k_{CO} is small Fig.11d, the lattice is still mostly filled with CO but the presence of O is more evident and there are almost no impurities, but when k_{CO} is large, Fig.11c, the lattice is mostly filled with O, some impurities, and the number of empty sites is slightly larger. Comparing Fig.11a with Fig.11c and Fig.11b with Fig.11d we see that the coverage of impurities does not appear to be very affected by the changes in p, but increases slightly with k_{CO}. Clearly when $p \leq 1$ most of the empty sites left from the desorption of CO are filled with O. This behavior is expected, since the O can not desorb and when p is very reduced they can hardly react, so they must remain on the surface. These plots illustrate the behavior already observed in the coverages and production plots.

4. Conclusions

In this study we performed Kinetic Monte Carlo simulations to include new effects into the ZGB model for the catalytic reduction of CO on a square lattice. In the original ZGB model a gas phase that contains CO and O_2, is in contact with a surface that can adsorb CO with probability y and O_2 with $1 - y$. Once adsorbed nn CO and O (product of the dissociation of the O_2 upon adsorption) react liberating CO_2. The ZGB model has two phase transitions, a continuous one at y_1 and a discontinuous one at $y_2 > y_1$. This very simple system not only qualitatively reproduce some basic aspects of the real catalytic process, but has become a very fruitful toy model for studying phase transitions in out of equilibrium systems. Experimentally it has been observed that temperature effects, that can be included

Journal of Physics: Conference Series **3052** (2025) 012004 doi:10.1088/1742-6596/3052/1/012004

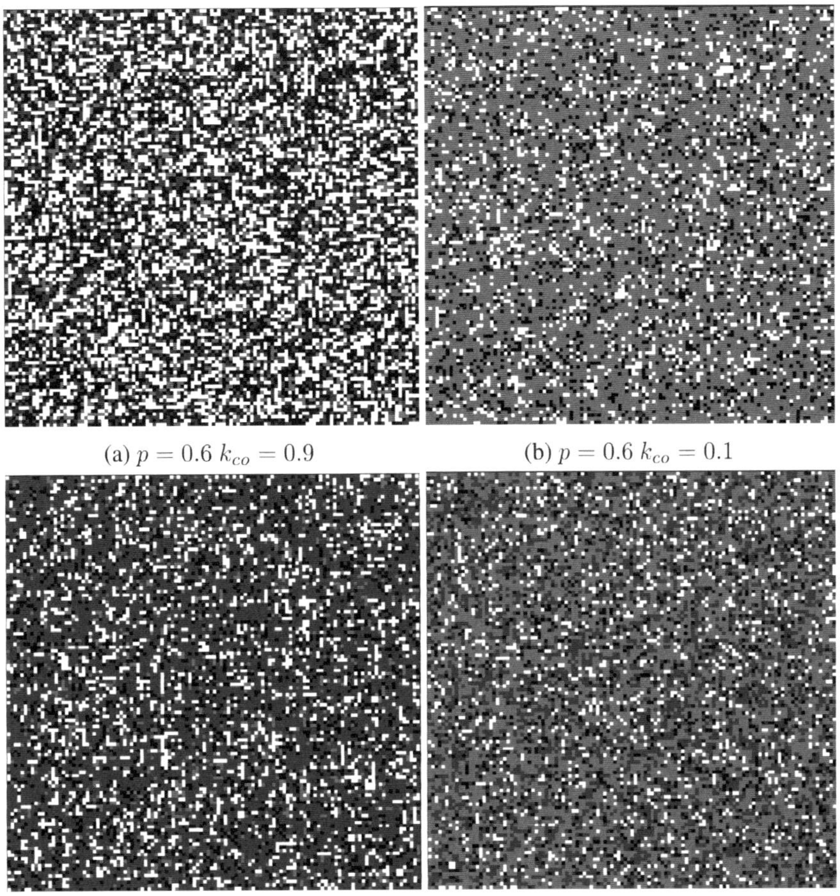

(a) $p = 0.6$ $k_{co} = 0.9$ (b) $p = 0.6$ $k_{co} = 0.1$

(c) $p = 0.1$ $k_{co} = 0.9$ (d) $p = 0.1$ $k_{co} = 0.1$

Figure 11: Snapshots of the lattice once the stable state has been reached for different values of k_{CO} and p with $y_X = k_X = 0.005$, and $y = 0.6$. ZGB+ model. The blue, red, black and white dots represent O, CO, X, and empty sites, respectively.

in the system as desorption rates of the adsorbates and changes in the rate of reaction, combined with the unavoidable existence of impurities in real systems, play a relevant role in the catalytic mechanism. Several authors have already introduced some of these effects into the model, in most cases separately. Our model, that we called ZGB+, takes into account the simultaneous effect of desorption of some of the adsorbates, a variable reaction rate, and impurities in the atmosphere. While the original model has only one parameter y, our model has four more; k_{CO} desorption probability of CO, $p \leq 1$ variable probability of $CO - O$ reaction, y_X associated with the existence of impurities in atmosphere, and k_X desorption rate of the impurities.

For comparison purposes we started studying separately the cases with k_{CO} and p variable, before adding the effects of the impurities. As other authors already found, the desorption of CO is relevant in the region $y \geq y_2$, where it has the effect of greatly increase the productivity and the range of values of y for which the system remains productive [19, 29]. The reduction of the reaction probability, $p \leq 1$, has the opposite effect, as p decreases the productivity drops abruptly at lower values of $y \leq y_2$ and the peak of the productivity also decreases until for a small enough value of p it disappears. When the original feeding gas includes impurities that can be adsorbed on the surface, it is found that if they can not desorb the system does not have a stable productive phase: eventually it reach an adsorbing state filled with X

Journal of Physics: Conference Series **3052** (2025) 012004 doi:10.1088/1742-6596/3052/1/012004

and CO. If we include the desorption of CO, we found that when k_{CO} is large it can take a considerable amount of time to reach this non-productive state, during which the system has some productivity. If the impurities can desorb, a stable reactive region reappears and becomes larger as k_X increases. All these factors have the effect of smoothing the abrupt phase transition at y_2 that presents the original ZBG model.

The model that includes all the above effects, the ZGB+, behaves quite differently from the previous ones. It is interesting how some effects compensate each other, the existence of impurities and the lower rate of reaction, p, reduces the productivity, while the desorption of CO and X increase it. The combined effect is the disappearing of the abrupt transition at y_2, making the changes in the coverages smoother, and increasing the production window. In general we appreciate that the effect of CO desorption is more relevant in the region of large y, while the effect of the reduced p and the desorption of impurities are more relevant in the region of lower y. If the number of impurities is low compared with their desorption rate they become irrelevant. In all the cases if the impurities do not desorb the system does not have a stable productive state. The desorption of CO only delays this process, and the reduced value of p accelerated it, but the final result is the same. The combination of effects alters the coverages and the productivity of the system in the whole range of values of y. We hope that these results can shed some light on the behavior of catalytic systems under different conditions, particularly temperature changes, and can inspire the design of changes that help to optimize these important processes.

References

[1] Armor J N 2011 *Catalysis Today* **3** 3–9
[2] Erisman J W, Sutton M, Galloway J, Klimont Z and Winiwarter W 2008 *Nat. Geosci.* **1** 636–9
[3] Hanefeld U and Lefferts L 2018 *Catalysis: An Integrated Textbook for Students* (Weinheim: Wiley-VCH)
[4] Ziff R M, Gulari E and Barshad Y 1986 *Phys. Rev. Lett.* **56** 2553–6
[5] Chang R and Goldsby K A 2016 *Chemistry* (New York: McGraw-Hill Education)
[6] Meakin P and Scalapino D J 1987 *J. Chem. Phys.* **87** 731–41
[7] Machado E, Buendía G M and Rikvold P A 2005 *Phys. Rev. E.* **71** 031603–11
[8] de Andrade M F and Figueiredo W 2023 *Physics Letters A* **475** 128863
[9] de Andrade M F and Figueiredo W 2016 *Physics Letters A* **380** 2628–31
[10] Fernandes H A and da Silva R 2019 *Journal of Statistical Mechanics: Theory and Experiment* **2019** 053205–17
[11] Buendía G M and Rikvold P A 2015 *Physica A Stat. Mech. Appl.* **424** 217–24
[12] Chan C H and Rikvold P A 2015 *Phys. Rev. E* **91** 012103–13
[13] Santos E D, Ferreira L S, Jorge L N, Leão S A and Caparica A A 2022 *Phys. A: Stat. Mech. Appl.* **589** 126663
[14] Cheimarios N 2023 *Model. Simul. Mat. Sci. Eng.* **31** 065007
[15] Fichthorn K A, Gulari E and Ziff R M 1991 *Surface Science* **243** 273–94
[16] Butt J B and Petersen E E 1989 *Activation, Deactivation, and Poisoning of Catalysts* (San Diego: Academic Press)
[17] Baron K 1978 *Thin Solid Films* **55** 449–62
[18] Bartholomew C H 2001 *Appl. Catal. A Gen.* **212** 17–60
[19] Ehsasi M, Matloch M, Franck O, Block J H, Christmann K, Rys F S and Hirschwald W 1989 *J. Chem. Phys.* **91** 4949–60
[20] Guo X and Yates J T 1989 *J. Chem. Phys.* **90** 6761–6
[21] Gland J L and Kollin E B 1983 *J. Chem. Phys.* **78** 963–74
[22] Machado E and Buendía G M 2003 *Phys. Rev. B* **68** 224411–7
[23] Ojeda C and Buendia G M 2012 *J. Comp. Meth. Sci. Eng.* **12** 261–7
[24] Wintterlin J, Schuster R and Ertl G 1996 *Phys. Rev. Lett.* **77** 123–6
[25] Jansen A P J 2012 *An Introduction to Kinetic Monte Carlo Simulations of Surface Reactions* (Heidelberg: Springer Berlin)
[26] Laidler K J, Meiser J H and Sanctuary B C 1999 *Physical Chemistry* (New York: Houghton Mifflin Company) p 857
[27] Buendía G M, Machado E, and Rikvold P A 2009 *J. Chem. Phys.* **131** 184704–6
[28] Fernandes H A, da Silva R and Bernardi A B 2018 *Phys. Rev. E* **98** 032113–9
[29] Buendía G M and Rikvold P A 2013 *Phys.Rev.E.* **88** 012132–6

Journal of Physics: Conference Series **3052** (2025) 012005 doi:10.1088/1742-6596/3052/1/012005

Synthesis and characterization of silver clusters deposited by magnetron sputtering on graphene systems

J A Castañeda-Vía[1†], L Akashi[2], M Yactayo[3,4], C V Landauro[3,4], J Quispe-Marcatoma[3,4] and A Champi[2†]

[1] Escuela de Posgrado "Víctor Alzamora Castro", Universidad Peruana Cayetano Heredia, Av. Honorio Delgado 430, Lima 15102, Peru.
[2] Laboratório de Novos Materiais de Carbono: Grafeno, Universidade Federal do ABC, Santo Andre 09210-170, Brazil.
[3] Facultad de Ciencias Físicas, Universidad Nacional Mayor de San Marcos, Calle Germán Amézaga 375, Lima 15081, Peru.
[4] Centro de Investigaciones, Tecnológicas, Biomédicas y Medioambientales (CITBM), Calle José Santos Chocano 199, Bellavista, Callao, Peru.

†Correspondence e-mails: jose.castaneda@upch.pe, ana.champi@ufabc.edu.br

Abstract. In this work, we show graphene bilayer and graphene multilayer (MLG) systems on which Ag clusters were deposited by magnetron sputtering. Atomic force microscopy shows that these clusters are organized in sizes of 200, 280 and 310 nm when deposited on SiO_2/Si substrates for short deposition times of 6, 10 and 14 seconds, respectively. On the other hand, the Ag clusters increase their sizes when deposited on graphene systems to same times and deposition conditions. The formation of structural disorder due to Ag clusters on the surface of graphene bilayers is clearly observed through Raman spectroscopy. This is evident by a strong increase in the intensity of the D and D' bands when compared to MLG with Ag clusters.

1. Introduction

Due to its large surface area and facility to anchor metal nanoparticles or clusters, graphene is used to form nanocomposites, which can extend its range of applications. Research has shown that graphene-metal composites have different applications in many fields such as fuel cells, optics, supercapacitors, etc. [1] The incorporation of silver into graphene has potential applications in antibacterial activity, catalysis, biosensors, etc. [2] Furthermore, silver nanoparticles into graphene will modify Raman peaks improving the signal, producing the SERS (surface enhanced Raman scattering) effect. [3,4]

Silver ultra-thin films deposited by magnetron sputtering, or other chemical or physical techniques, tend to produce large grain sizes, and high surface roughness due to 3D island growth (Volmer-Weber growth mode). [5] Thus, the silver thin film growth starts with nucleation due to the high free surface energy. [6] It is important to set up the correct working pressure if it is required to control the sample growing. [5]

In this context, in the present work we study the synthesis of silver clusters (Ag-clusters) on graphitic systems (bilayer and MLG) by magnetron sputtering. The analysis of their structural properties is also presented. The work is organized as follows: In the next section we present the experimental procedure

Journal of Physics: Conference Series **3052** (2025) 012005 doi:10.1088/1742-6596/3052/1/012005

carried out for the synthesis of silver clusters on graphitic systems. Then, we present the discussion of the results and the summary of this work.

2. Experimental procedure

SiO_2/Si substrates were pretreated using the procedure described by Akashi. [7] This process consists of a sequence of cleaning with isopropyl alcohol, heating at low temperature and cleaning with ultrasound, as shown in Figure 1.

Multilayer graphene (MLG) was obtained by mechanical exfoliation of graphite flake (Nacional de Grafite, Minas Gerais, Brazil) and then deposited on SiO2/Si substrates. [8] Using an optical microscope, the number of layers of MLG can be observed. [9]

Silver clusters deposition was performed in an ATC Polaris magnetron sputtering system (AJA International Inc., Massachusetts, USA), with an Argon flow of 20 sccm and 20 W of source power with a nominal deposition rate of 0.57 Å/s. The substrates both with and without MLG were placed on a sample holder (Figure 1) with a rotation of 20 rpm at 20 cm from the silver target. The working pressure used was 4.0 x 10^{-3} Torr, during deposition time of 6, 10 and 14 seconds.

Atomic force microscopy measurements were made with an Alpha300 RA microscope (WITec GmbH, Ulm, Germany). Contact mode micrographs were taken at 256 pixels x 256 pixels on a 15 μm x 15 μm surface. The set-point voltage chosen for the cantilever was 0.4 V, and a scanning speed of 1 second per line. The magnetic ring with the cantilever was coupled to a Zeiss EC Epiplan 20x optical objective to perform the measurements.

Raman spectroscopy measurements were carried out by using an InVia Reflex spectrometer (Renishaw, Wotton-under-Edge, UK) using an Argon laser whose wavelength is 514 nm used as a source of excitation. The spectra were recorded in the spectral region from 1200 cm^{-1} to 3000 cm^{-1} with a step of 1.2 cm^{-1}.

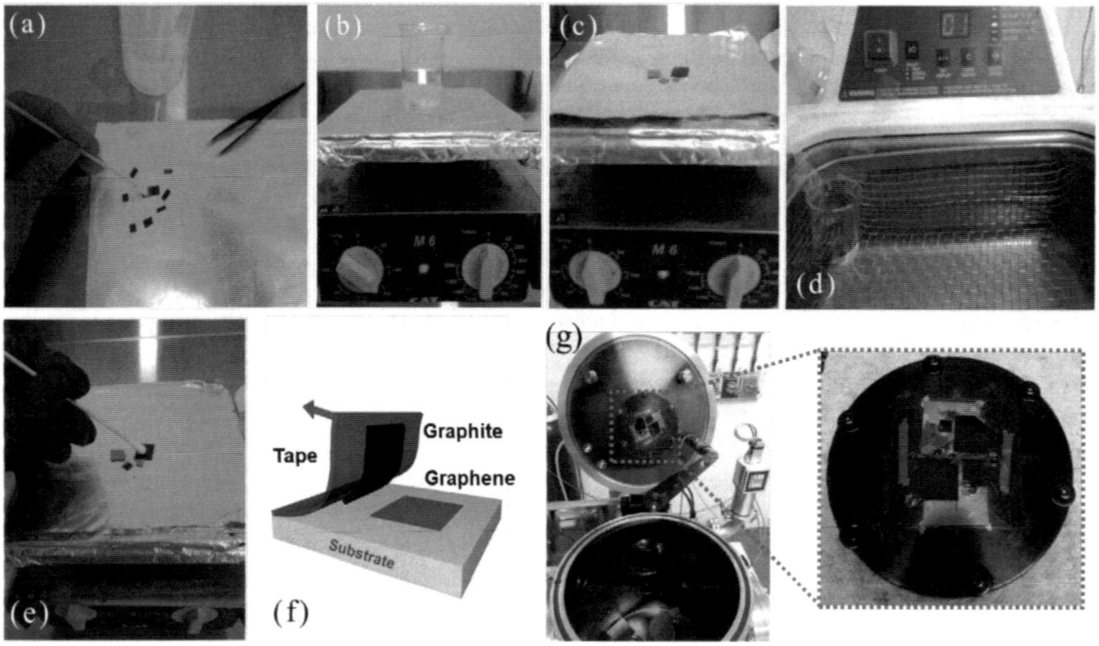

Figure 1. (a-e) Substrate cleaning procedure. (f) Schematic representation of the micromechanical exfoliation process. (g) Substrates with and without MLG placed on the sample holder into magnetron sputtering chamber.

3. Results and discussions

Journal of Physics: Conference Series **3052** (2025) 012005
doi:10.1088/1742-6596/3052/1/012005

The deposition of Ag-clusters on the substrates with and without MLG forms small silver agglomerates which are more distinguishable as the deposition time increases, as observed in Figure 2. The presence of stacked layers of graphene can also be observed according to the difference in light contrast under the light microscope.

Figure 2. Optical micrographs of Ag-clusters deposited for (a) 6 seconds and (b) 10 seconds with MLG, exfoliated on SiO_2/Si substrates.

Atomic force microscopy measurements allowed us to observe the size of the deposited Ag-clusters and the height of the MLG at the nanometric scale (Figures 3 - 8). In the samples with MLG (Figures 3, 5 and 7) it was possible to determine that the thickness of the graphene layers is approximately 1,5 nm, which is equivalent to a graphene bilayer. [10]

The Ag-clusters deposited on SiO_2/Si substrates, without MLG, have average diameters of 200, 280 and 310 nm for deposition times of 6, 10 and 14 seconds, respectively (Figures 4, 6 and 8). On the other hand, Ag-clusters deposited on MLG have average diameters of 620, 700 and 740 nm for the same deposition times (Figures 3, 5 and 7). It can be seen that Ag-clusters grow more easily on MLG than on SiO_2/Si substrate. [11]

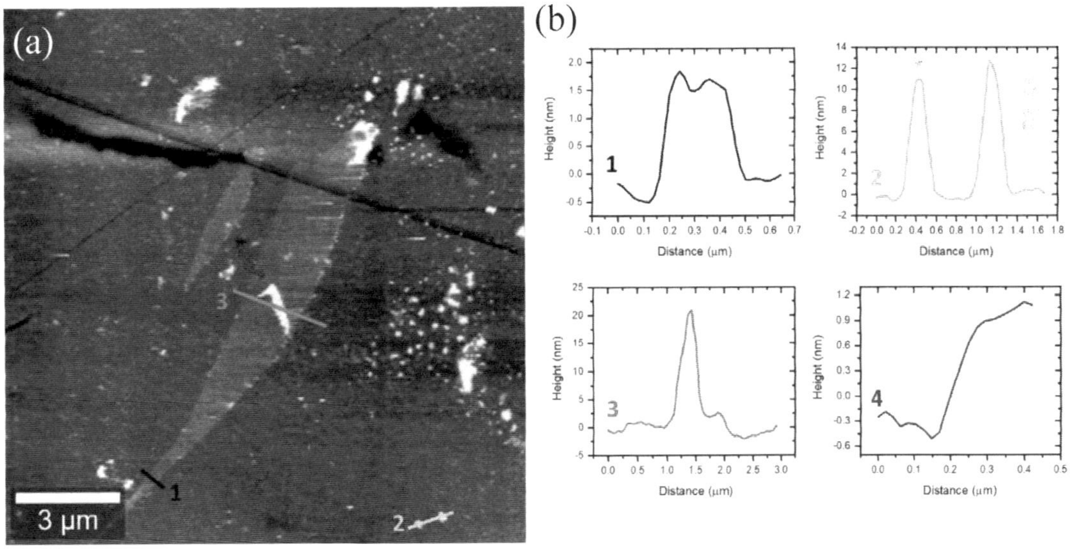

Figure 3. Atomic force micrograph of Ag-clusters deposited for 6 seconds on SiO_2/Si substrate containing MLG. (a) AFM micrograph and (b) its linear profiles.

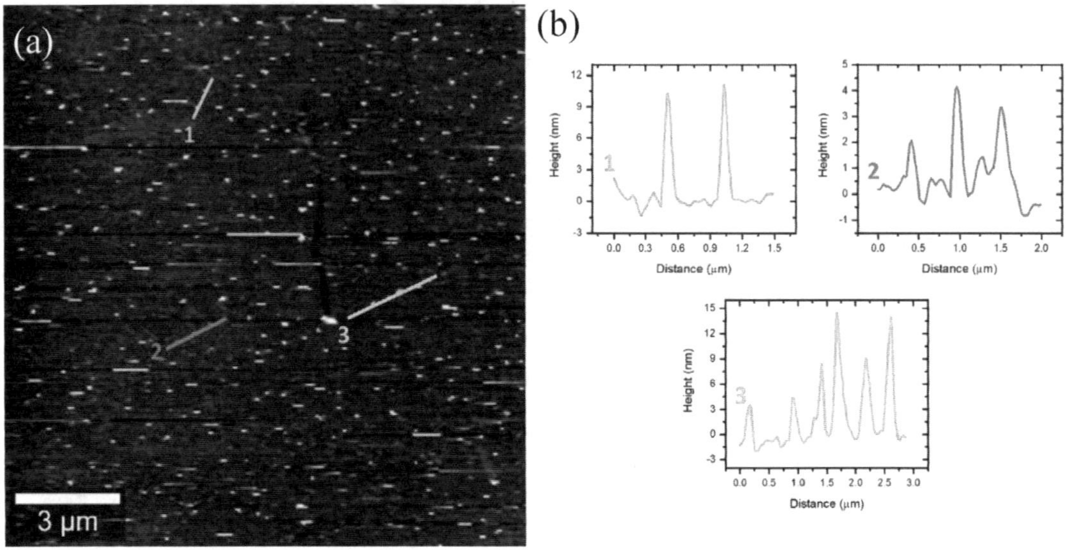

Figure 4. Atomic force micrograph of Ag-clusters deposited for 6 seconds on SiO_2/Si substrate without MLG. (a) AFM micrograph and (b) its linear profiles.

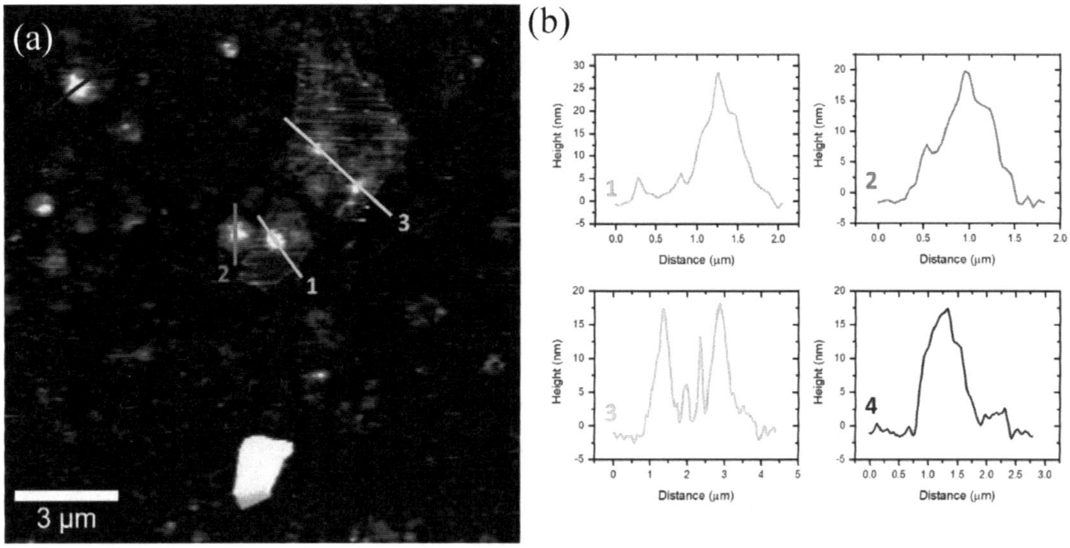

Figure 5. Atomic force micrograph of Ag-clusters deposited for 10 seconds on SiO_2/Si substrate containing MLG. (a) AFM micrograph and (b) its linear profiles.

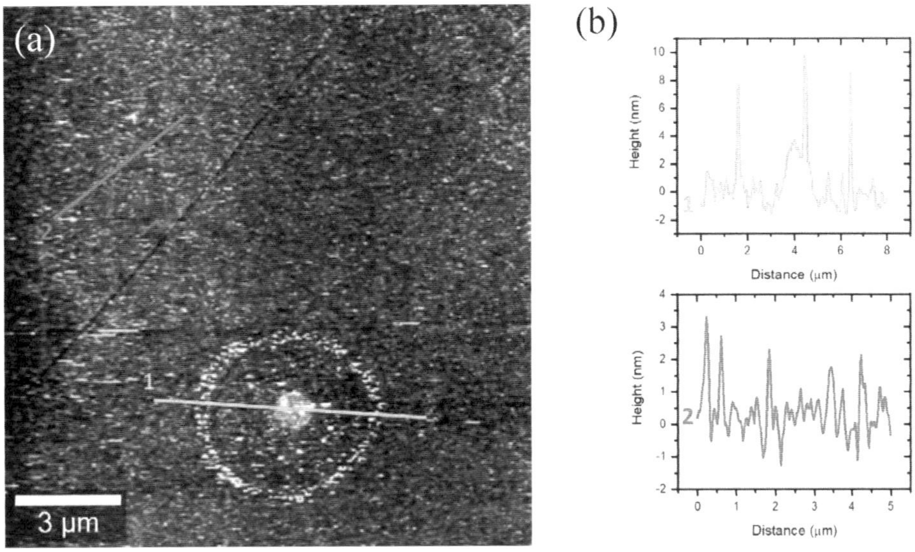

Figure 6. Atomic force micrograph of Ag-clusters deposited for 10 seconds on SiO_2/Si substrate without MLG. (a) AFM micrograph and (b) its linear profiles.

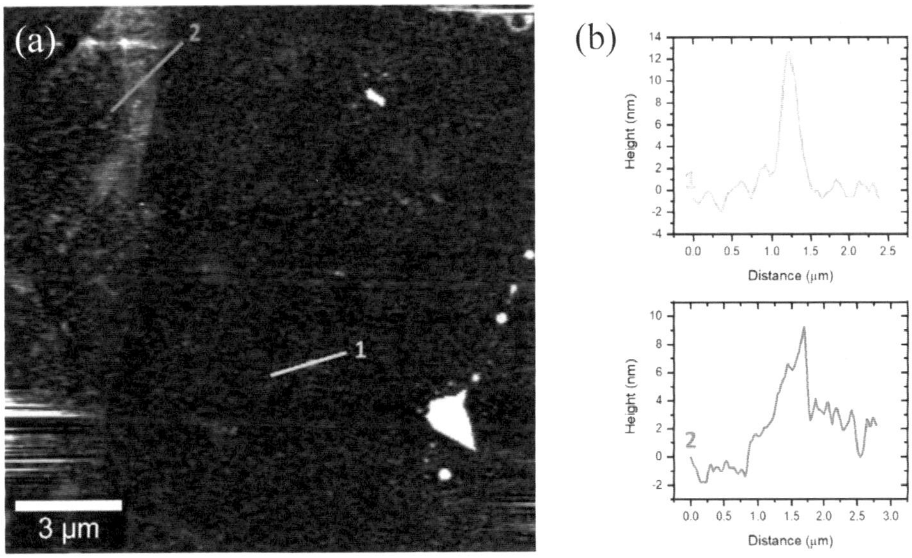

Figure 7. Atomic force micrograph of Ag-clusters deposited for 14 seconds on SiO_2/Si substrate containing MLG. (a) AFM micrograph and (b) its linear profiles.

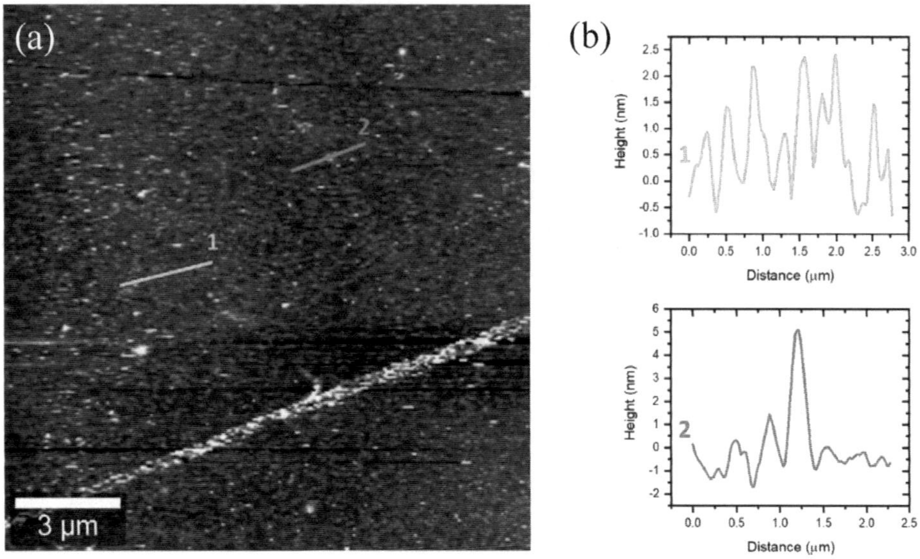

Figure 8. Atomic force micrograph of Ag-clusters deposited for 14 seconds on SiO_2/Si substrate without MLG. (a) AFM micrograph and (b) its linear profiles.

To quantitatively determine the effect of MLG on the size of the deposited Ag-clusters, the diameter values were fitted with a linear curve to find the growth rate of the clusters, being the values found 15.0 ± 2.9 nm/s and 13.8 ± 3.6 nm/s for the cases with and without MLG, respectively (see Figure 9). These results do not show significant differences, so the diameters only have a constant difference of approximately 420 nm.

Figure 10 shows the typical Raman spectra corresponding to a graphene bilayer (Point 1) and MLG (Point 2), in agreement with the atomic force microscopy observations. Figure 10a, shows that when there are more layers of graphene, the G band around 1600 cm-1 has higher intensity and the 2D band has more components, while the presence of the D band is not observed. [10] On the other hand, Figure 10b, shows the spectra with the deposited Ag-clusters, in which the presence of the D band can be observed. The D band is associated with the structural disorder of the graphene layers, which indicates that the graphene bilayer is more susceptible to deformation in the presence of the Ag-clusters and consequently the D band has greater intensity.

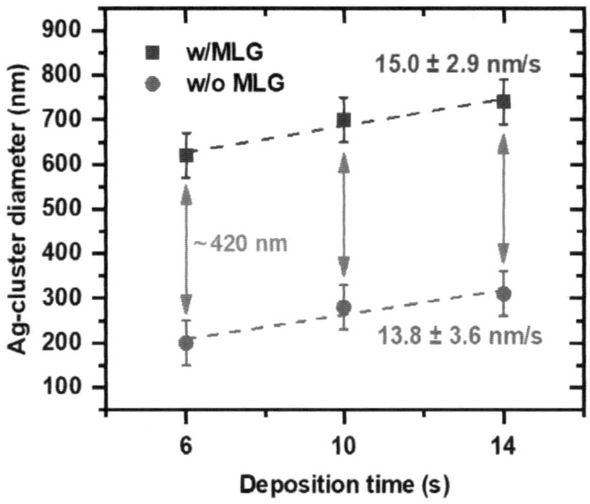

Figure 9. Diameters of Ag-clusters deposited on SiO_2/Si substrate without MLG (green circles) and containing MLG (blue squares) for each time of deposition.

Journal of Physics: Conference Series **3052** (2025) 012005
doi:10.1088/1742-6596/3052/1/012005

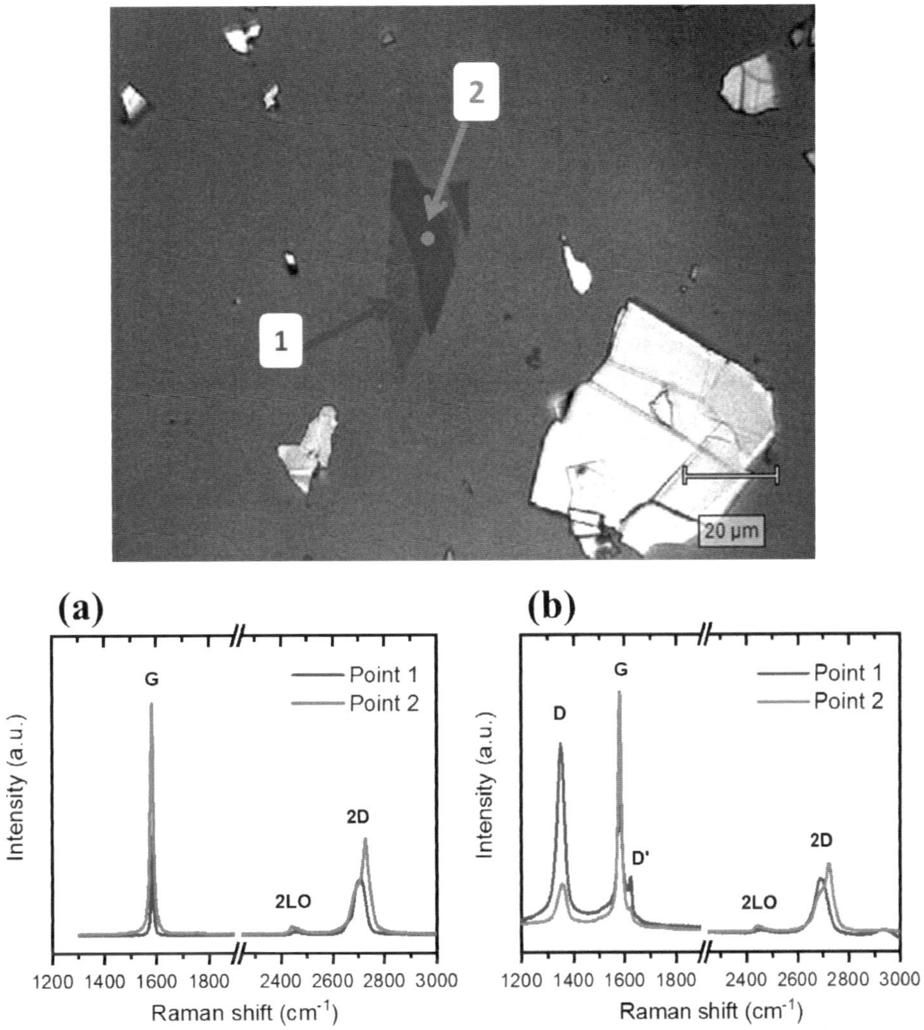

Figure 10. Raman spectra of the graphitic system (points 1 and 2 of the micrograph, respectively) (a) without Ag-clusters deposition and (b) with Ag-clusters deposited for 14 seconds.

4. Conclusions

The current investigation demonstrates the feasibility of growing Ag-clusters on MLG. To achieve this, we initially employed the micromechanical exfoliation technique to deposit MLG on SiO_2/Si substrates, followed by the utilization of magnetron sputtering to grow Ag-clusters. The size of the silver clusters can be controlled by deposition time. In fact, to avoid the formation of an Ag-layer we employ short deposition times, up to 14 seconds, such that only Ag clusters are formed, as can be observed in the atomic force microscopy images. The deposition of Ag-clusters on graphene bilayer and MLG induced structural disorder, which is evident from the appearance of the D band in both systems observed by Raman spectroscopy. The higher intensity observed in the graphene bilayer system, compared to the multilayer system, suggests the presence of the D' band at approximately 1630 cm^{-1}.

Acknowledgments

The authors are grateful to Universidad Nacional Mayor de San Marcos for partial financial support by contract R.R. N° 03556-R-19 under project number B19130021. Also, the authors are grateful to Dr. Jorge Andrés Guerra from the Department of Physics of the Pontificia Universidad Católica del Perú

Journal of Physics: Conference Series **3052** (2025) 012005 doi:10.1088/1742-6596/3052/1/012005

(PUCP) for the Raman spectroscopy measurements, and Dr. Gretty Villena and Dr. Ilanit Samolski from the Laboratorio de Micología y Biotecnología of the Universidad Nacional Agraria La Molina (UNALM) for the atomic force microscopy measurements. J. C.-V. recognizes the financial support of the CONCYTEC-World Bank Project "Improvement and Expansion of the Services of the National System of Science, Technology and Technological Innovation" 8682-PE, through its executing unit FONDECYT (Grant number 08-2018). C. V. L. and J. Q.-M. are grateful to Consejo Nacional de Ciencia, Tecnología e Innovación Tecnológica, Republic of Peru, for partial financial support through the Excellence Centers Program. A. Ch. would like to thank for the financial support to "National Council for Scientific and Technological Development – CNPq". Process: 408397/2022–5, FAPESP Grant (2022/11894-1) and CNPq process: 302584/2023-4. Finally, the authors are very grateful to SBF, in particular to the Latin American Physics Program (PLAF) for funding the exchange of students between both Peru and Brazil.

References

[1] Su H and Hu Y H 2021 *Energy Sci. Eng.* **9** 958
[2] He K, Zeng Z T, Chen A W, Zeng G M, Xiao R, Xu P, Huang Z Z, Shi J B, Hu L and Chen G Q 2018 *Small* **14** 1800871
[3] Liu J, Liu L, Wu X, Zhang X and Li T 2015 *New J Chem* **39** 5272
[4] Sukumaran S S, Rekha C R, Resmi A N, Jinesh K B and Gopchandran K G 2018 *J. Sci.: Adv. Mater. Devices.* **3** 353
[5] Devaraj V, Lee J, Baek J and Lee D 2016 *Appl. Sci. Converg. Technol.* **25** 32
[6] Goetz S, Bauch M, Dimopoulos T and Trassl S 2020 *Nanoscale Adv.* **2** 869
[7] Akashi L 2016 *Estudo da influência do substrato na obtenção de novos materiais de Grafeno - Scientific Initiation Thesis* (Santo Andre: Federal University of ABC)
[8] Ciric L, Sienkiewicz A, Nafradi B, Mionic M, Magrez A and Forro L 2009 *Phys. Status Solidi* B **246** 2558
[9] Obelenis F and Champi A 2014 *Braz. J. Phys.* **44** 682
[10] Li D, Zhou Y S, Huang X, Jiang L, Silvain J-F and Lu Y 2015 *Nanoscale* **7** 3651
[11] Deng D, Feng S, Shi M and Huang C 2017 *J. Mater. Sci.: Mater. Electron.* **28** 15411
[12] Ferrari A 2007 *Solid State Commun.* **143** 47

Journal of Physics: Conference Series **3052** (2025) 012006 doi:10.1088/1742-6596/3052/1/012006

Why pseudo-complex General Relativity? and Applications

Peter O. Hess

Instituto de Ciencias Nucleares, Universidad Nacional Autónoma de México, Circuito Exterior, C.U., A.P. 70-543, 04510 México D.F., Mexico
and
Frankfurt Institute for Advanced Studies, Johann Wolfgang Goethe Universität, Ruth-Moufang-Str. 1, 60438 Frankfurt am Main, Germany

E-mail: hess@nucleares.unam.mx

Abstract. A brief discussion on the *pseudo-complex General Relativity* is presented. It is shown that this theory is a viable extension of GR, with deviations centered near to the event horizon. The theory introduces a dark energy accumulation, due to the coupling to the central mass. Predictions of this theory are resumed, as for example the structure in an accretion disk, with a dark ring followed by a bright ring further in. The current *Event Horizon Telescope* observation of M87 is not able to discriminate between GR and pcGR, due to a low resolution. Further predictions are also discussed, as the physics of neutron stars, the redshift at the surface of the star and *Quasi Periodic Object*.

1. Introduction

General Relativity (GR) is one of the best tested theories in existence. Solar system experiments [1] confirm the theory, as recent observations of gravitational waves [2] and the observation of a black hole shadow by the *Event Horizon Telescope* (EHT) [3].

Nevertheless, one has to keep testing a theory at its extreme limits, as for very strong gravitational fields. Though, the EHT observation is consistent with GR, its resolution is not sufficient to discriminate theories which only deviate from GR near the event horizon. Further, there are conceptual problems in the case of a black hole, as the singularity in its center, the information loss and the existence of an event horizon which separates the inner region from the outer one. The existence of an event horizon is a matter of opinion. The event horizon is just a coordinate singularity and there exist others, well established ones. However, this one is the result of a strong gravitational fields and it may bother that the interior of a black hole "in the corner of a room" cannot be accessed.

One proposal to extend GR is the *pseudo-complex General Relativity* (pcGR) which makes definite predictions for the region near the event horizon [4, 5, 6, 7] and also avoids the event horizon. The theory has no singularity and information can get out of the black hole, because of that rather called a *black star*, though, in many respects this dark star behaves like a black hole. Deviations from GR are only noticeable near the event horizon. Less massive object, as neutron stars, behave in pcGR as they do in GR.

In this contribution, I present the motivation for the pcGR and why it is a viable proposal for an extension of GR. Also, some of the predictions made by this theory will be resumed.

2. An algebraic extension of General Relativity

Earlier attempts on extending GR algebraically have been proposed by A. Einstein [8, 9] and M. Born [10, 11]. Einstein extended the metric to a complex one, where the real part is the metric used in GR and the imaginary component is associated to the electromagnetic energy-momentum tensor. The motivation was to unify GR with Electrodynamics, but unfortunately in the limit of small curvature the theory of Electrodynamics was not recovered and this approach was abandoned. This extension is equivalent to introducing complex coordinates instead of x^μ [7]. M. Born's motivation was different: He noted that coordinates and momenta are not treated equivalently, as is done in Quantum Mechanics. He introduced the concept of *complementary* and added a metric term quadratic in the momenta. In order to observe units, the momentum term carries a dependence on a *minimal length scale factor*. This renders integrals over momenta finite, another reason why M. Born advocated this extension. The appearance of a minimal length scale, as a parameter, has also the advantage that Lorentz symmetry is not broken. Later, E.R. Caianiello [12] modified M. Born's length element and the effects of the minimal length were discussed in [13].

All these extensions can be summarized under *algebraic extensions*. In an algebraic extension the coordinates are redefined as

$$
\begin{aligned}
x^\mu \quad \rightarrow \quad X^\mu &= x^\mu + \sum_k a_k y_k^\mu \\
a_k a_p &= \sum_q C_{kpq} a_q \quad .
\end{aligned}
\tag{1}
$$

The algebraic relation in the second line in (1) is the reason to call it an *algebraic extension*. The meaning of y_k^μ has to be deduced for each type of extension.

The simplest example is the *complex extension* $X^\mu = x^\mu + iy^\mu$, with $i^2 = -1$, as proposed in [14]. Another, not so well known extension, is to *pseudo-complex coordinates* (pc) (also called hyperbolic coordinates) $X^\mu = x^\mu + Iy^\mu$, with $I^2 = +1$.

There are many more possible coordinate systems, but the important point is that in [15] all possible algebraic extensions were investigated if they contain ghost and/or tachyon solutions, considered as rendering a theory inconsistent. They found that only two algebraic versions of the coordinates are allowed, namely real coordinates, leading to GR, and pseudo-complex coordinates, leading to a theory called the *pseudo-complex General Relativity*. The main path in [15] is to consider the weak field limit and determine the propagators of gravitational waves. In this approach and restricting here to the complex or pseudo-complex extension only, two types of propagators appear, one with a factor of 1 and another with i^2 or I^2, for the complex and pseudo-complex case respectively. In the complex case $i^2 = -1$ and refers thus to a ghost solution. Only in the pseudo-complex case the factor I^2 is still 1, rendering this extension as the only viable one.

For this reason, the pcGR was considered as a viable extension to GR. In pcGR the infinitesimal length square element has the same form as in GR, but now in terms of the pc-variables:

$$
\begin{aligned}
d\omega^2 &= g_{\mu\nu}(X) dX^\mu dX^\nu \\
g_{\mu\nu} &= g_{\mu\nu}^R + I g_{\mu\nu}^I \ , \ X^\mu = x^\mu + Iy^\mu \quad .
\end{aligned}
\tag{2}
$$

with, this the length element acquires the form

$$
d\omega^2 = \left\{ g_{\mu\nu}^R \left[dx^\mu dx^\nu + dy^\mu dy^\nu \right] + g_{\mu\nu}^I \left[dx^\mu dy^\nu + dy^\mu dx^\nu \right] \right\}
$$

$$+I\left\{g^I_{\mu\nu}\left[dx^\mu dx^\nu + dy^\mu dy^\nu\right] + g^R_{\mu\nu}\left[dx^\mu dy^\nu + dy^\mu dx^\nu\right]\right\} \quad . \tag{3}$$

Concerning the real part, all formerly mentioned extensions, mentioned above, can be accommodated, i.e., it also contains a minimal length. Because a particle can move only along a real path, the pseudo-imaginary component of the length element has to vanish, which leads to the *constraint*

$$g^I_{\mu\nu}\left[dx^\mu dx^\nu + dy^\mu dy^\nu\right] + g^R_{\mu\nu}\left[dx^\mu dy^\nu + dy^\mu dx^\nu\right] \ = \ 0 \quad . \tag{4}$$

To solve (4) is particular easy in a flat space, with $g^R_{\mu\nu} = \eta_{\mu\nu}$ and $g^I_{\mu\nu} = 0$. The constraint then reduces to (skipping a factor of 2)

$$\eta_{\mu\nu}dx^\mu dy^\mu \ = \ 0 \quad , \tag{5}$$

which is nothing but the standard dispersion relation with the solution $y^\mu \sim u^\mu$, i.e., y^μ is proportional to the 4-velocity. For dimensional reasons, a minimal length scale factor has to be introduced, and using $c = 1$ the y^μ acquires the form

$$y^\mu \ = \ lu^\mu \quad . \tag{6}$$

Thus, the y^μ components are related to the minimal length scale. In this case the length element of E.R. Caianiello can be recovered. A more general, approximate solution of (4) is given in [7].

In order to obtain the extended Einstein equations, one uses the standard form of the action

$$S = \int dX^4 \sqrt{-g}\left(\mathcal{R} + 2\alpha\right) \quad , \tag{7}$$

where $\mathcal{R}$ is the pc-Riemann scalar and α is related to the dark, energy as in GR. For cosmological models, the α has to be a constant due to translational invariance. However, in a central problem the α may depend on r and with rotation included also on the azimuthal angle θ. In contrast to GR, all elements in (7) are pseudo-complex. For a mathematical introduction on the analysis of pc-variables and for further references, please consult [5].

The action principle $\delta S = 0$ is applied. In former publication a modified variational principle was proposed. But as shown already on the last pages of [5], including the constraint (4) it is equivalent to use the standard variational principle. *Neglecting the minimal length* finally leads to the Einstein equations

$$\mathcal{R}_{\mu\nu} - \frac{1}{2}g_{\mu\nu}\mathcal{R} \ = \ 8\pi T^\Lambda_{\mu\nu} \quad . \tag{8}$$

For more details, please consult [7]. In a current investigation [16], the influence of a minimal length within pcGR is investigated, with the result that only for masses of the order of l appreciable changes occur, thus justifying to neglect l for macroscopic black holes. The energy-momentum tensor depends on a couple of parameters which leaves space for further assumptions. This leads to the introduction of dark energy, with the density ϱ_Λ.

In addition to the pseudo-complex structure, pcGR assumes that dark energy accumulates near a black hole (or a mass in general), but only noticeable near the event horizon. This is a result of semi-classical calculations in Quantum Mechanics [17, 18], i.e., dark energy is created in a curved space-time back ground. The pcGR assumes therefore that there is a coupling between the central mass and the amount of dark energy. The ansatz for the dark energy is

$$\varrho_\Lambda \sim \frac{B_n}{r^{n+2}} \quad , \tag{9}$$

with B_n and n being parameters of the theory.

This assumption implies an important principle, namely that *a central mass not only curves space.time but also changes the vacuum properties of space-time near it.* This coupling to the dark energy is a consequence of quantum effects and the study of pcGR therefore can help in understanding what these quantum effects might be. The other parameter is n, describing the fall-off of the dark energy as a function in r. In [4] $n = 2$ was assumed, but this already violates observations in the solar system. From that, calculations were performed with $n = 3$ [5], however, in [19, 20] the inspiral wave of a gravitational wave was fitted to pcGR and concluded that n has to be larger than 3. From then on, the value $n = 4$ is assumed, which is also in better accordance to the r-dependence obtained in [18]. Nevertheless, one has to keep in mind that B_n and n are parameters, converting pcGR to a phenomenological theory. There exist studied for the accumulation of *dark matter* in a black hole. dark matter does not interact strongly with matter, only through gravitation. In contrast, dark energy does interact with matter [18] and should play a more dominant role.

The metric in pcGR, including rotation, has the form [7] (there is an error in the g_{11}-component, corrected here)

$$\begin{aligned}
g_{00} &= -\frac{r^2 - 2m_0 r + a^2\cos^2\vartheta + \frac{B_n}{(n-1)(n-2)r^{n-2}}}{r^2 + a^2\cos^2\vartheta} \quad , \\
g_{11} &= \frac{r^2 + a^2\cos^2\vartheta}{r^2 - 2m_0 r + a^2 + \frac{B_n}{(n-1)(n-2)r^{n-2}}} \quad , \\
g_{22} &= r^2 + a^2\cos^2\vartheta \quad , \\
g_{33} &= (r^2 + a^2)\sin^2\vartheta + \frac{a^2\sin^4\vartheta\left(2m_0 r - \frac{B_n}{(n-1)(n-2)r^{n-2}}\right)}{r^2 + a^2\cos^2\vartheta} \quad , \\
g_{03} &= \frac{-a\sin^2\vartheta\, 2m_0 r + a\frac{B_n}{(n-1)(n-2)r^{n-2}}\sin^2\vartheta}{r^2 + a^2\cos^2\vartheta} \quad .
\end{aligned} \tag{10}$$

It is illustrative to calculate the energy-momentum tensor within the pc-Schwarzschild limit (non-rotating star). Substituting the metric (10) into the left hand side of the Einstein equations results in the $T_{\mu\nu}$ on the right hand side of these equations. Further, for the energy-momentum tensor we define the dimensionless quantities $y = \frac{r}{m_0}$ and $b = \frac{B}{m_0^4}$, with which this tensor acquires the form (the signature of the metric is (-+++))

$$8\pi\left(T^\mu{}_\nu\right) = \begin{pmatrix} -\frac{b}{2y^6} & 0 & 0 & 0 \\ 0 & -\frac{b}{2y^6} & 0 & 0 \\ 0 & 0 & \frac{b}{y^6} & 0 \\ 0 & 0 & 0 & \frac{b}{y^6} \end{pmatrix} = 8\pi \begin{pmatrix} -\varrho^\Lambda & 0 & 0 & 0 \\ 0 & p_r^\Lambda & 0 & 0 \\ 0 & 0 & p_\vartheta^\Lambda & 0 \\ 0 & 0 & 0 & p_\vartheta^\Lambda \end{pmatrix} \quad . \tag{11}$$

From (11) the Riemann scalar $R = -\frac{b}{y^6}$ is deduced. It is clearly seen that the energy-momentum tensor describes an anisotropic fluid with a different pressure in the radial compared to the angular directions.

From these consideration, it should be clear that the pseudo-complex extension of GR has to be considered seriously. The next question is: What kind of prediction the pcGR is making and in what they differ to GR?

3. Resumé of some predictions of pcGR

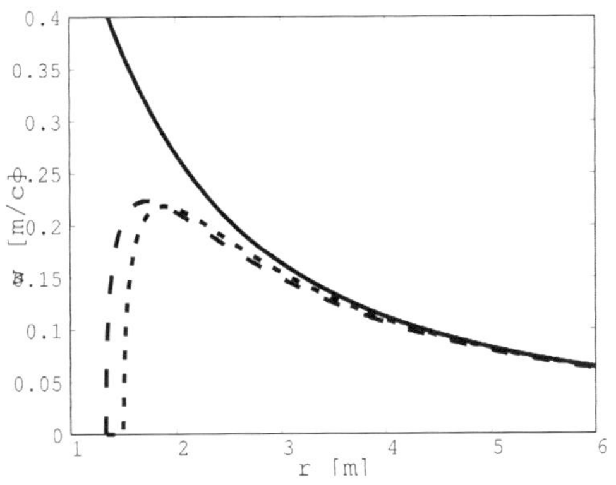

Figure 1. The orbital frequency of a point particle in a circular orbit as a function in r. The rotational parameter is $a = 0.95m_0$. The upper curve is the result as obtained in GR while the two lower curves are from pc-GR, for $n = 3$ (left curve) and $n = 4$ (right curve).

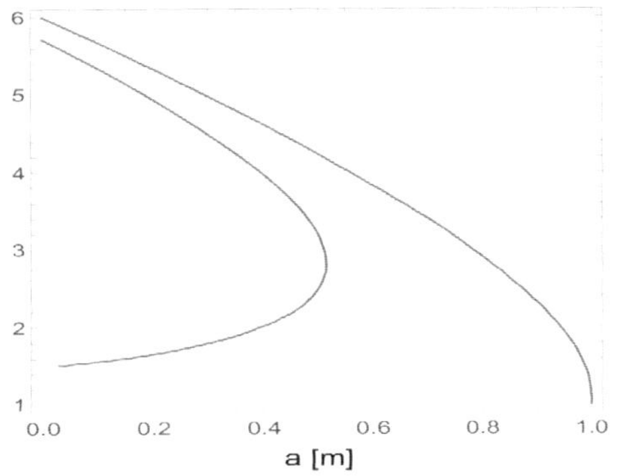

Figure 2. Limits for stable orbits ($n = 4$) as a function in the rotational parameter a. The upper curve is the result of GR, which marks the position of the *Innermost Stable Circular Orbit* (ISCO). In pc-GR there are no stable orbits to the left of the lower curve.

3.1. Structure of accretion disks

In [5, 6, 22, 23] simulations of a thin, optical thick accretion disk [21] where published and compared to GR.

In order to understand the results of the simulations, we have to explain how the orbital frequency of a particle in a circular orbit changes with respect to the radial distance and where do stable orbits exist: In Fig. 1 this orbital frequency is depicted for the Kerr parameter $a = 0.95m_0$. Clearly seen is the appearance of a maximum, at which two neighboring orbitals

Journal of Physics: Conference Series **3052** (2025) 012006 doi:10.1088/1742-6596/3052/1/012006

Figure 3. Simulations of the accretion disc in M87, using the model of Ref. [21] and $a = 0.95m_0$. The left panel is for a resolution of 5μas. The right panel is for 20μ as, a little bit less than 24μas, the resolution reported by the EHT. The inclination angle of the disc with respect to the observer is 70°.

have the same orbital frequency, which results in less friction and less emission if light, marking the position of the dark ring. In Fig. 2 the range of stable orbits is depicted [24]. The upper line corresponds to GR, where for $a = 0$ the last stable orbit is at $6m_0$ and it approaches $1m_0$ for $a = 1m_0$. The lower curve encircles the area where no stable orbit exists in pcGR. Note that above $a = 0.5m_0$ all orbits are stable in pcGR and they can reach the point where the orbital frequency has a maximum. Below $a = 0.5m_0$ the pcGR follows the GR 1, i.e., similar results are expected, though, the last stable orbit is further in in more light is emitted.

In Fig. 3, on the left panel the simulation for a high resolution is given. It is clearly observed that the disk has a dark ring, followed further in by a bright one. The dark ring is the consequence of the maximum in the orbital frequency as a function on the radial distance. The right panel in Fig. 3 shows the same but with a low resolution of 20μas, a bit lower than the EHT. One can see that the ring structure is washed out. Thus, in order to see the ring structure one has to increase the resolution of the EHT significantly. It also means that the observations are consistent with GR *and* pcGR. Thus, this particular prediction of pcGR cannot be verified yet and one has to await a much better resolution.

3.2. Redshift

There is hopefully a possibility to see some difference between GR and pcGR, but it depends on the non-existence of a jet. Because the jet is a consequence of the presence of an accretion disk, this is equivalent to demand that there is no accretion disk.

The only observable black hole up to now by the EHT, which may have no accretion disk, is SgrA* in the center of our Galaxy. Observational results are still pending.

In Fig. 4 the redshift *at the surface of the black star* is plotted as a function on the azimuthal angle θ. The different curves correspond to different a-values. The redshift is lowered toward the poles and if a is very large (above $0.5m_0$), the redhift has values of 1-2. In other words, infalling matter, which collides with the surface, should emit light and its reshifted frequency has to be detected by an observer on earth. Toward the orbital plane the redshift is too large

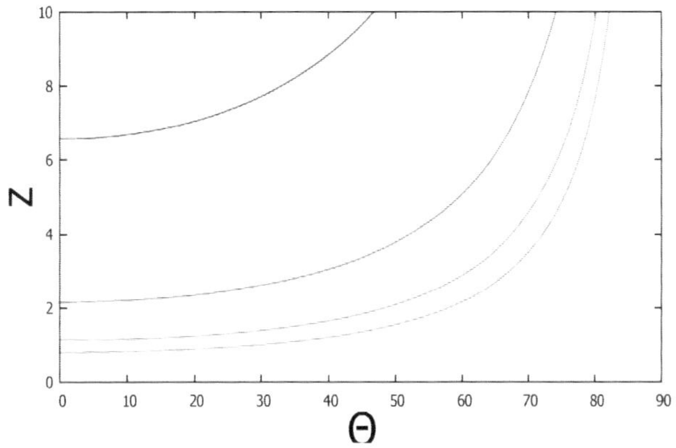

Figure 4. Redshift z at $r = \frac{3}{2}m_0$ for different rotation parameter values a, as a function on the azimuth angle θ. From the upper to the subsequent lower curves the a-value varies from $0.2m_0$, $0.5m_0$, $0.8m_0$ to $1.0m_0$.

and no hope exist to detect it. I.e., there is only hope to see something near the poles, which requires that, first, the matter has to fall onto the surface near the poles and, secondly, there is no jet present (due to an accretion disk), which would over-shine the effect.

3.3. Neutron Stars

In the first decade of this century a neutron star with 2.05 solar masses was observed [25], while until then one did not expect masses larger than 2, with a "realistic" equations of state. In [26] a theory for neutron stars was developed which finally could accommodate the mass. Recently, in observing gravitational waves [27], a further candidate with 2.5 solar masses was suggested. Since then, many contributions can be found which can explain these masses. The central point is how to change the equation of state of the matter in order to support a greater mass. This seems a bit unsatisfactory, because greater masses can obviously obtained by changing appropriately the equation of state, such that larger masses can be obtained.

In pcGR this is not a problem: The accumulation of dark energy, due to the coupling of the mass to it, accommodates well larger masses, as is shown in [5, 28]. There, the coupling of the mass density with the dark energy density was assumed to be linear

$$\varrho_\Lambda = \alpha \varrho_{\rm m} \quad . \tag{12}$$

The index m refers to the mass density and the index Λ to the dark energy density. Also here the mass couples to the dark energy. In fig. 5 several curves for different α-values are depicted. Stars up to 6 solar masses were found stable. However, if these dark stars can still be considered neutron stars is questionable. In our view, there is a continuous shift from a classical neutron star, with a particle beam, to a classical black hole, where the emitted light of a beam shifts to a large redshift or disappears due to unknown processes.

In [29] semi-classical calculations [17] were performed which resulted in a more complicated relation for the coupling of the matter density to the dark energy density. As a result, the coupling diminished toward to the surface of the black star and much higher masses were obtained.

All calculations relied on the hadronic model for the matter, published in [26]. With increasing density this model reaches its limit and due to lack of confidence no further calculations can be

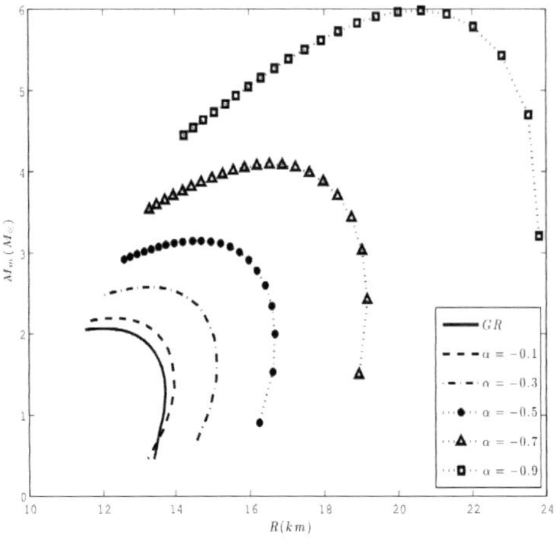

Figure 5. Mass vs the radius of a black star. A linear relation between the dark energy density and mass density in the interior of the star was assumed.

done. In this respect, new models for the matter at high density are required.

Therefore, we predict the existence of stellar objects which behave as a neutron star, emitting a beam whose emitted light has a large redshift, and also have too large masses to be explained by standard models.

3.4. Quasi Periodic Objects (QPO)

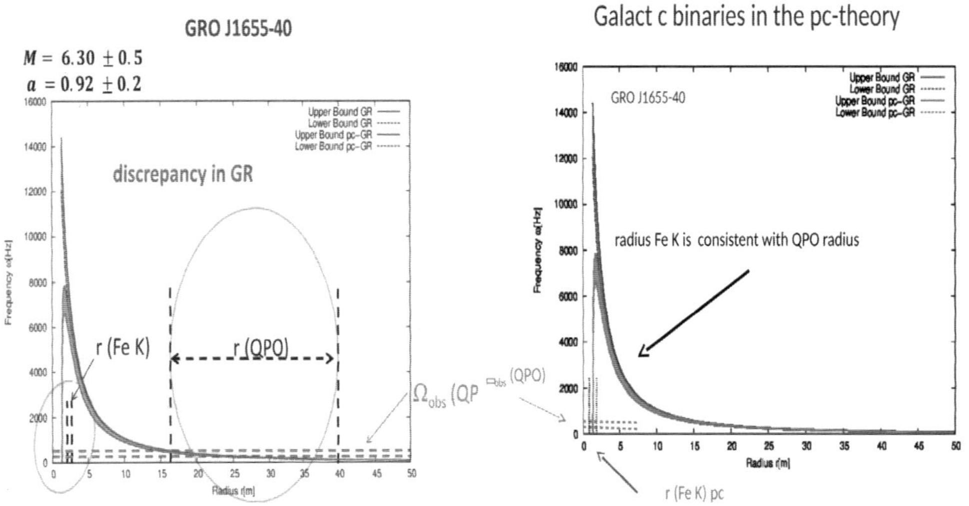

Figure 6. The orbital frequency of a particle in a circular orbit versus the radial distance. The upper curve is for GR while the lower one for pcGR and $n = 3$ (for $n = 4$ the result is not much different). The width of a line is a measure of the error in the a-parameter. The horizontal dashed line is from fitting to the orbital frequency observed, with its experimental error. The vertical dashed line is for adjusting to the redshift of the emitted line.

QPO's are assumed to be excitations within an accretion disks (hot spots) which are circling around with the disk, They are observed in accretion disks of galactic black holes and also in the disks of stellar black holes [30, 31, 32, 33]. Knowing the frequency and the formula for the orbital frequency in a circular orbit, a distance value can be deduced. If this is the real distance has to be verified by another measurement, as the redshift of the line emitted. Unfortunately, for galactic black holes this has not been seen yet. Because galactic black holes are fairly well isolated (no large partner nearby), the interpretation of the QPO as a co-rotating hot spot in the accretion disk is quite well established. For stellar black holes one also observes QPO's and simultaneously the rotation frequency and the redshift of the emitted line. Therefore, from the redshift one can deduce also a distance. *From both observations the same distance has to result*, for consistency. In Fig. 6 the orbital frequency of a particle in a circular orbit is depicted. The upper line refers to GR while the lower line to pcGR. The horizontal dashed line is from adjusting to the observed orbital frequency and the vertical line from the observed redshift of the emitted line.

The problem is now, using the formulas from GR both distances are *not the same*, but they *are the same* in pcGR. One might say that this proves pcGR, however, one has to take into account possible effects of the stellar partner onto the accretion disk around the black hole. It was shown that one can indeed explain the discrepancy in GR by this influence of the stellar partner, using models [34]. Though pcGR is easier to apply here, not involving further assumptions, the fact that one can reconcile the measurement with GR makes it inconclusive.

Conclusions

I have presented the motivation for the algebraic extension of GR to pseudo-complex coordinates. The pcGR turns out to be the *only viable algebraic extension.*

The structure of pcGR was explained and its differences to GR pointed out.

A theory has to be verifiable, which is exactly the case for pcGR. Several predictions were presented:

- The appearance of a dark ring followed further in by a bright ring, which is due to the dependence of the orbital frequency of a particle in a circular orbit. Unfortunately, the resolution of the EHT is not sufficient to distinguish pcGR from GR, until a much better resolution is reached.
- The redshift of light emitted from the surface of the dark star, as a function of the azimuthal angle was determined. We predict that matter falling onto the surface near the poles can emit light with a redshift between 1-2, which should be observable. However, this is linked to the condition that the massive object is not accompanied by an accretion disk, which would include a jet near the poles. This jet would over-shine the effect described. The black hole in the center of our galaxy, SgrA*, is thought not to have an accretion disk, so there is some hope.
- The theory allows a stable star for any mass value, i.e., it easily accommodates larger masses for neutron stars, without any further assumptions, using an equation of state known. Possibly, there is a smooth transition from stellar object which behave as a neutron star with a beam, to stellar objects where the beamdisappears. The uncertainty here is the structure (equation of state) within the star.
- Quasi Periodic Objects, which are observed in galactic centers and in binary systems with one black hole. Here, the simultaneous measurement of the orbital frequency and the redshift of the emitted line lead to *the same distance*, contrary to GR. However, investigating the effect of the stellar visible partner on the structure of the accretion disk, using models, can still accommodate GR. This is a case where pcGR leads to a simple explanation, but still the situation is not settled yet.

Journal of Physics: Conference Series **3052** (2025) 012006 doi:10.1088/1742-6596/3052/1/012006

As can be seen, the pcGR is a realistic extension of GR, but further observations with their improvements have to be made.

Acknowledgment
Financial support from DGAPA-PAPIIT (IN100421 is acknowledged.

References

[1] C. M. Will, Living Rev. Relativ. **9** (2006), 3.
[2] Abbott B. P. et al. (LIGO Scientific Collaboration and The Virgo Collaboration), 2016c, Phys. Rev. Lett. **116**, 241103.
[3] The Event Horizon Telescope collaboration, ApJ **875** (2019), L1.
[4] P. O.Hess and W. Greiner, Int. J. Mod. Phys. E **18**, 51 (2009).
[5] Hess P. O., Schäfer M. and Greiner W., *Pseudo-Complex General Relativity*; Springer: Heidelberg, Germany, 2015.
[6] P. O. Hess and E. López-Moreno, Universe **5** (2019), 191.
[7] P. O. Hess, Progr. Part. Nucl. Phys. 114 (2020), 103809.
[8] A. Einstein, Ann. Math. **46**, 578 (1945).
[9] A. Einstein, Rev. Mod. Phys. **20**. 35 (1948).
[10] M. Born, Proc. Roy. Soc. A **165**, 291 (1938).
[11] M. Born, Rev. Mod. Phys. **21**, 463 (1949).
[12] E. R. Caianiello, Nuovo Cim. Lett. **32** (1981), 65.
[13] A. Feoli G. Lambiase, G. Papini and G. Scarpetta, Phys. Lett. A **263** (1999), 147.
[14] C. L. M. Mantz and T. Prokopex, arXiv:gr-qc.0804.0213
[15] P. F. Kelly and R. B. Mann, Class. and Quant. Grav. **3**, 705 (1986).
[16] L. Maghlaoui and P. O. Hess, manuscript in preparation (2022).
[17] N. D. Birrell and P. C. W. Davies, Quantum Fields in Curved Space (Cambridge University Press, The Edinbugh Building, Cambridge, 1994).
[18] M. Visser, Phys. Rev. D **54** (1996) 5116.
[19] A. Nielsen and O. Birnholz, Astron. Nachr. **339** (2018), 298.
[20] A. Nielsen and O. Birnholz, Astron. Nachr. **340** (2019), 116.
[21] D. N. Page and K. S. Thorne, Astrophys. J. **191** (1974), 499.
[22] T. Boller, P. O. Hess, A. Müller ands H. Stöcker, MNRAS:Letters **485** (2019), L34.
[23] P. O. Hess, T. Boller, A. Müller and H. Stöcker, MNRAS:Letters **485** (2019), L121.
[24] T. Schönenbach, G. Caspar, P. O. Hess, T. Boller, A. Müller and W. Greiner, Mon. Not. R. Astron. Soc. **442** (2014),121–130.
[25] D. J. Nice, E. M. Splaver, I. H. Stairs, O. Loehmer, A. Jessner, M. Kramer and J. M. Cordes, ApJ **634** (2005), 1242.
[26] V. Dexheimer and S. Schramm, The Astrophys. J. **683** (2008) 943.
[27] R. Abbott, T. D. Abbott, F. Acernese, et al., arXiv:2111.03606[gr-qc]; Report No.: LIGO-P200031
[28] I. Rodríguez, P. O. Hess, S. Schramm and W. Greiner, J. Phys. G **41**, (2014), 105201.
[29] G. Caspar, I. Rodríguez, P. O. Hess and W. Greiner, Int. J. Mod. Phys. E **25** (2016), 1650027.
[30] T. H. Belloni, A. Sanna and M. Mendz, MNRAS **426** (2012), 1701.
[31] R. T. Hynes, D. Steeghs, J. Casares, P. A. Charles and K. O'Brian, The Astrophys. J. **609** (2004), 317.
[32] R. C. Reis, A. Fabian, R. R. Ross, G. Miniutti, J. M. Miller and C. Reynolds, MNRAS **387** (2008), 1489.
[33] J. Steiner, J. McClintock, G. Jeffrey et al., 38th COSPAR Scientific Assembly, 18–15 July, Bremen, Germany, 2010).
[34] D. Lai, W. Fu, D. Tsang, J. Horak and C. Ya, Proc. of the Intern. Astron. Union **290** (2012), 57.

AUTHOR INDEX